Foreword

The Information Systems Engineering Library provides guidance on managing and carrying out Information Systems Engineering activities. In the IS life cycle, Information Systems Engineering takes place once the IS strategy has been defined. It is concerned with the development and ongoing improvement of information systems to the operational stage and their maintenance whilst in operational use.

The Information Systems Engineering Library complements other CCTA products, in particular the project management method, PRINCE and the systems analysis and design method, SSADM.

Volumes in the Information Systems Engineering Library are of interest to varying levels of staff from IS directors to IS providers, helping them to improve the quality and productivity of their IS development work. Some volumes in this library should also be of interest to business managers, IS users and those involved in market testing, whose business operations depend on having effective IS support by means of Information Systems Engineering activities.

The Information Systems Engineering Library also complements other related CCTA publications, particularly the IT Infrastructure Library for operational issues, the IS Planning Subject Guides for strategic issues, the RISK Library for the identification and assessment of risks, and the Programme and Project Management Library which covers issues relating to the effective management of programmes and projects.

CCTA welcomes customer views on Information Systems Engineering Library publications. Please send your comments to:

Customer Services
Information Systems Engineering Group
Rosebery Court
St Andrew's Business Park
Norwich NR7 0HS

Contents

Chapter		Page
1	**Introduction**	9
	1.1 Purpose	
	1.2 Who should read this volume	
	1.3 The structure of this volume	
	1.4 The goals of this volume	
	1.5 Terminology	
2	**Overview**	13
	2.1 Risk reduction in IS acquisition	
	2.2 Customer and supplier responsibilities and roles	
	2.3 Specifying the requirement	
	2.4 SSADM and acceptance testing	
	2.5 SSADM and system maintenance	
	2.6 SSADM user skills	
3	**Procurement and the development life cycle**	21
	3.1 The development life cycle	
	3.2 Procurement stages in the life cycle	
	3.3 Producing the SOR and OR	
	3.4 Management decision points	
	3.5 Procuring high complexity systems	
4	**SSADM in the procurement process**	33
	4.1 The role of SSADM in the procurement process	
	4.2 Procurement and the 3-schema specification architecture	
	4.3 SSADM in systems procurements	
	4.4 SSADM in services procurements	
	4.5 Benefits from the use of SSADM	
	4.6 Limitations in the use of SSADM	
5	**The Feasibility Study**	47
	5.1 Role of the Feasibility Study	
	5.2 Procurement issues for the study	
	5.3 The SSADM products	
	5.4 Supporting non-SSADM products	
	5.5 Documenting the options	
	5.6 Identifying Full Study and procurement activities	

6		**The Full Study**	**65**
	6.1	Role of the Full Study	
	6.2	SSADM stages for a procurement Full Study	
	6.3	Requirements and background information	
	6.4	The Statement of Requirement (SOR)	
	6.5	The Operational Requirement (OR)	
7		**The System Procurement Study (SPS)**	**105**
	7.1	The role of the SPS	
	7.2	SSADM stages in the SPS	
	7.3	SSADM products from the SPS	
	7.4	Evaluation of the SPS deliverables	
8		**After Award of Contract**	**127**
	8.1	Outline	
	8.2	Monitoring system development	
	8.3	Acceptance Trials	
	8.4	Post Implementation Review	
	8.5	Systems support	
9		**The Requirements Catalogue to the SOR/OR**	**135**
	9.1	Formatting the Requirements Catalogue	
	9.2	Level of requirement detail	
	9.3	Defining mandatory and desirable requirements	
10		**SSADM products and acceptance testing**	**149**
	10.1	The role of the Acceptance Trial	
	10.2	SSADM traceability paths	
	10.3	SSADM products in acceptance testing	
11		**SSADM products and system maintenance**	**175**
	11.1	Contracted-out and in-house support	

12	**Phased procurements**	**191**
	12.1 The purpose of phased procurements	
	12.2 The impact of phasing on SSADM	
13	**The market testing context**	**203**
	13.1 The impact on system development	
	13.2 New systems procurement	
	13.3 Responsibilities for system support	
	Bibliography	**207**
	Glossary	**209**
	Index	**215**

VOLUME STRUCTURE

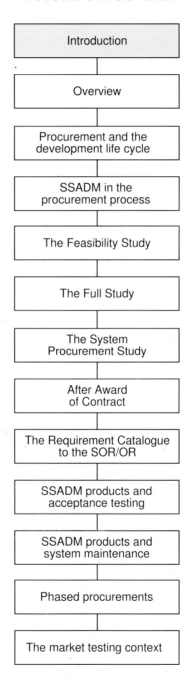

1 Introduction

1.1 Purpose

The Structured Systems Analysis and Design Method (SSADM) is a systematic approach to the analysis and design of IT applications, originated and promoted by CCTA. It is the method recommended for use in UK Government Departments. Many IT applications are procured from external suppliers. This volume provides definitions and guidance on practical interfaces between analysis and design activities and products, and the procurement process and products.

SSADM can make a direct and valuable contribution to the procurement process. As an analysis and design method SSADM has an obvious role in the procurement of systems that include a significant amount of bespoke development, but SSADM may also be used in a number of other types of system procurements. It may also be of benefit to some types of support services procurement.

The SSADM manuals describe how to specify a requirement for an Information System (IS) where the focus of the method is on a seamless move from the specification of the requirement into the design of the system. Procuring a part or the whole of a system from an external supplier introduces a break in this otherwise seamless transition.

The volume seeks to help the customer to effectively manage this break by providing a definition of the possible interfaces between SSADM and IS procurement.

References to procurement are made in general terms wherever possible. Where detailed descriptions of the procurement process are necessary, the methodology of the Total Acquisition Process (TAP) is used. TAP is the Government's recommended approach for IS procurement, where a system requirement needs to be specified, and the value of the system and supporting services is above the EC/GATT threshold.

Where procurements are subject to EC Supplies Directive or EC Services Directive, care will be needed to avoid breaching regulations which may constrain the specification of the use of SSADM by the supplier. Expert advice should always be sought in case of doubt.

This volume is drafted to conform to the high-level requirements of Euromethod, but specific Euromethod terminology is rarely used to avoid possible misunderstanding by the intended readership.

1.2 Who should read this volume

The volume is intended to be used by project managers who are responsible for procurements in which SSADM is used to communicate between customer, supplier and SSADM practitioners working on procurement projects. Suppliers of IS/IT systems competing for procurements in which SSADM is used will also find it of value.

1.3 Assumed knowledge

It is assumed that readers are familiar with and have free use of SSADM. Those not familiar with SSADM terminology may have to consult with SSADM experts or refer to the SSADM Version 4 manuals.

1.4 The structure of this volume

The volume adopts a top-down structure to the role of SSADM in procurement, summarized in the diagram preceding this chapter. This diagram is repeated at the beginning of each following chapter, to aid navigation through the volume. To enhance its readability the volume structure maps to the project life cycle, with two final chapters addressing more general subjects.

Within the readership outlined in Section 1.2, some may require a detailed description of how SSADM should be used on a procurement project, while others may only need to appreciate the central issues. To assist the latter, each section commences with a highlighted paragraph summarizing the text that follows.

An overview of the volume is given in Chapter 2. The following chapters (3 and 4) set the context for procurement in the information systems development life cycle, and SSADM in the procurement process.

Chapters 5 to 8 then describe in detail the use of SSADM in procurement projects, in the sequence of the development life cycle. The Feasibility Study is discussed in Chapter 5, and the Full Study in Chapter 6. Chapter 7 describes the use of SSADM, by both the customer and

Chapter 1
Introduction

the supplier, during the System Procurement Study (SPS), and Chapter 8 considers the stages after the award of contract.

The volume then moves to a further level of detail. Chapter 9 describes how to prepare the Requirements Catalogue so that the requirements might be in a suitable format for 'pasting' into the Statement of Requirement (SOR) and Operational Requirement (OR), the two key procurement documents. SSADM's relevance to acceptance trials is described in Chapter 10, including the use of requirements' traceability paths and the SSADM specification products which support them. Chapter 11 suggests the set of SSADM products with which maintenance of the system can best be supported.

A phased procurement strategy is often adopted for high risk procurements, and Chapter 12 describes the use of SSADM in this approach. Finally, Chapter 13 discusses the additional factors that need to be considered when a Department's IT services have been market tested, and particularly when those services are outsourced.

1.5 The goals of this volume

The four goals that have guided the development of this volume are as follows:

- to explain the impact on SSADM when the method is used to support the procurement of an information system (as opposed to the development of the system 'in-house')

- to distinguish the SSADM activities and products which are the responsibility of the customer from those which are the responsibility of the supplier

- to demonstrate how SSADM can be used as a direct input to the preparation of procurement products

- to explain the constraints which the procurement process may place on the use of SSADM.

1.6 Terminology

Throughout the volume, SSADM terminology is used when referring to systems specification and design processes, and TAP terminology is used in respect of the procurement processes.

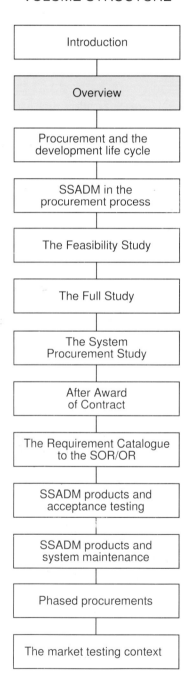

2 Overview

2.1 Risk reduction in IS acquisition

Both SSADM and good procurement practice are concerned with the reduction of risk and the support of success in an IS project. In this overview chapter, the ways in which this is achieved are reviewed. Particular attention is paid to customer and supplier responsibilities, the contribution of SSADM to requirements specification and acceptance testing in the procurement process, and the requirements for products to support ongoing system maintenance.

Most of the concerns of the decision maker in the customer organization at the start of a system development project are likely to be rooted in risks, of one form or another. These include, amongst others:

- technical risks: for example, if adequate IT skills and resources are not available

- risks associated with evolving requirements

- risks associated with managing change in the user environment

- cost risks

- schedule risks.

Some of these risks may be reduced by seeking to procure a solution to the IS requirement rather than develop the system in-house. Also, in some circumstances, procurement from an external supplier may be the only practical option available. This volume assumes that the reader has already taken the decision to procure part or all of the IS.

Risks may be further reduced by the adoption of standard methods, such as:

- PRINCE, for the management and control of projects

- CRAMM, for the analysis and management of security risks

- SSADM, for the analysis and design of the system.

This volume assumes that the reader is considering, or has already decided, to use SSADM and that they are familiar with the basic concepts of SSADM.

SSADM pre-supposes a seamless transition from the analysis and specification of the requirement into the design of the solution. However, the decision to procure a solution introduces a break between the Requirements Specification, which is the responsibility of the customer, and the design of the IS, which is normally prepared by the supplier. Another break typically occurs between the design by the supplier and the acceptance trials conducted, or controlled, by the customer. The way in which SSADM is applied to a procurement project needs to take these breaks into account, to provide the smoothest possible interface and a contribution to the cost-effective management of the procurement process.

The structural model of SSADM is unaffected by the procurement, except that Stage 4, Technical System Options (TSO), is typically supplanted by the procurement process. On procurement projects the consideration of TSOs becomes, in practice, an iterative process, commencing in many instances with initial informal assessment of suppliers' likely solutions, and continuing with short-listing from among the Mini-Proposals (suppliers' responses to the Statement of Requirement) and Full Proposals (suppliers' responses to the Operational Requirement). The process finishes with the effective selection of a TSO when the implementation contract is awarded.

Most of the SSADM techniques are applicable on a procurement project for the requirement needs to be specified with precision, though not so detailed as to unnecessarily constrain the supplier. The rigour of the techniques in SSADM can deliver the required precision. The design by the supplier also needs to be complete, derived from the specification and testable against it, for all of which the SSADM design techniques are invaluable.

2.2 Customer and supplier responsibilities and roles

In a procurement project the customer and supplier roles, and the relationship of design activities and procurement activities within both customer and supplier must be clearly understood.

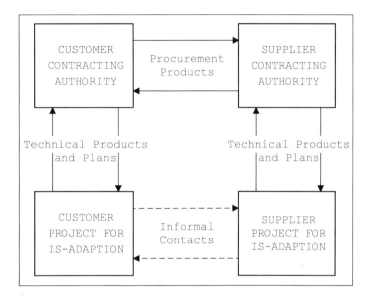

Figure 2.1: Roles and activities in a procurement project

These roles are shown graphically in Figure 2.1. This, exceptionally, employs Euromethod terminology to clearly differentiate between activities.
The customer is responsible for:

- Stage 0 Feasibility

- Stage 1 Investigation of Current Environment

- Stage 2 Business System Options

- Stage 3 Definition of Requirements

- Stage 4 (which becomes a procurement stage)

- (optionally) Stage 5 Logical Design.

The supplier is responsible for:

- (optionally) Stage 5 Logical Design

- Stage 6 Physical Design.

The responsibility for carrying out Stage 5 Logical Design should be decided on a case by case basis. There are circumstances when it is better for the customer to complete Stage 5, and circumstances when it is better for the supplier to complete it. Chapters 6 and 7 of this volume discuss this in detail.

2.3 Specifying the requirement

Amongst the objectives of this volume are, firstly, to describe to the customer how SSADM may be used in preparing the procurement Requirements Specification on a project. Secondly, to provide pointers for the benefit of both customer and supplier to the best use of SSADM in the design stage.

The way in which the customer uses SSADM on a procurement project must avoid duplication of effort in the preparation of the Requirements Specification and in the preparation of the procurement products. A secondary objective of this volume is to describe how the requirements sections of the two key procurement products, the OR and the SOR, can consist of requirements 'pasted' in from the SSADM Requirements Catalogue. Chapter 9 discusses this in detail.

The main focus of this volume is on fully bespoke solutions to the requirement but, where appropriate, references are made to how SSADM can be used in the procurement of application packages. The ISE Library volume: *SSADM and Application Packages* provides more detailed guidance on the use of SSADM in the procurement of application packages.

In specifying the requirement there is a balance to be struck by the customer. It is important that all of the requirement is described, and that it is specified to the level of detail needed by the supplier. If requirements are not specified or if they are specified at too low a level of detail, then there may be differences of opinion between the customer and the supplier about the supplier's responsibilities under the contract, possibly resulting in additional charges being incurred by the customer. On the other hand to specify the requirement to too high a level of detail may unnecessarily constrain a solution which the supplier might otherwise propose, to the detriment of the success of the IS implementation.

These issues are brought sharply into focus in a procurement project. System requirements in the SOR or OR should be labelled as 'mandatory' or 'desirable'. At the stage of proposal evaluation a proposal should be rejected if it does not meet all of the mandatory requirements. Also, if key requirements have been omitted from the Requirements Specification, the customer could incur additional charges to have them implemented later.

Services might also be procured at the same time as a system, and the same issues apply:

- how to be sure the supplier has understood the requirement

- how to test that the requirement has been met.

In order to be of help to the greatest possible number of managers, this volume considers all of these issues.

In terms of the second objective included at the beginning of this section, to provide pointers to the best use of SSADM in the design stage, SSADM does not prescribe the format of the physical design products. Therefore, the Stage 6 products required by the customer must be defined by the customer and supplier in agreement, since not all Stage 6 products will be appropriate in all cases. Once they have been defined, however, it is the responsibility of the supplier to propose the format of the products, taking the strengths and weaknesses of the development environment into account.

2.4	**SSADM and acceptance testing**	Within a procurement project, the question of whether or not a supplier has fulfilled his contractual obligations is to a great extent answered during acceptance testing by the customer. This volume describes (in Chapter 10) the requirements traceability paths which SSADM products support, and gives guidance on how to use SSADM products to 'drive' acceptance testing.
2.5	**SSADM and system maintenance**	Whether the system is to be maintained, in the short term, by the supplier, the customer or by a third party, it is important that the supplier, who designs the system, hands over the SSADM (or other) products by which enhancements to the system can be economically managed. These products need to be specified in the OR, and possibly in the SOR, and this volume suggests the set of SSADM products required for the purpose.
2.6	**SSADM user skills**	In the simplest terms, the successful development of an information system always depends upon the skills of the people involved – in the ability of the end-user to agree and express the requirement, and in the ability of the analysts, designers, builders and testers to develop a system which meets the requirement.

Neither the use of SSADM, nor the use of established procurement practices guarantees success. However, this volume seeks to bridge the 'culture gap' between SSADM users and procurement specialists, so that the two disciplines applied in harmony may have a greater opportunity to deliver successful information systems.

Chapter 2
Overview

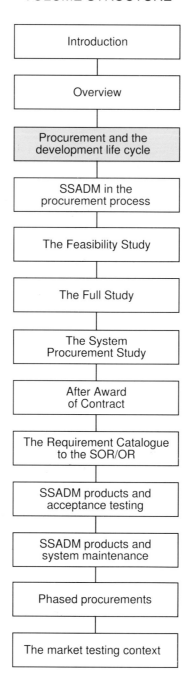

3 Procurement and the development life cycle

3.1 The development life cycle

Before considering the use of SSADM in procurement projects, it is important to understand exactly where SSADM and the procurement process fit into the development life cycle. The figures in this chapter illustrate these relationships graphically.

Figure 3.1 summarizes the IS development life cycle as a series of activities in the chronological order in which they happen. These activities are presented as a 'V'. Each development activity moving down the left side of the 'V' is concerned with a level of detail lower than the activity before it; each testing activity on the right side of the 'V' is concerned with a level of testing higher than the activity before it.

Figure 3.2 illustrates how the SSADM activities and the procurement activities are mingled together and shows all these activities in the chronological order in which they will normally happen.

3.2 Procurement stages in the life cycle

The procurement activities are also shown in Figure 3.1, and are summarized as a series of boxes along the bottom of the diagram, concluding with an open-ended box representing the procurement process beyond award of contract, which also includes the acceptance testing. This section discusses the interaction between SSADM and procurement activities.

In Figure 3.1, the development activities are shown coinciding with the linked procurement activities. Given the range of procurements to which this diagram applies, it necessarily gives a simplified view of the links. But the diagram does illustrate two key issues:

- inputs to the procurement process can start early in the life cycle

- Stage 4 of SSADM is largely supplanted by the procurement process.

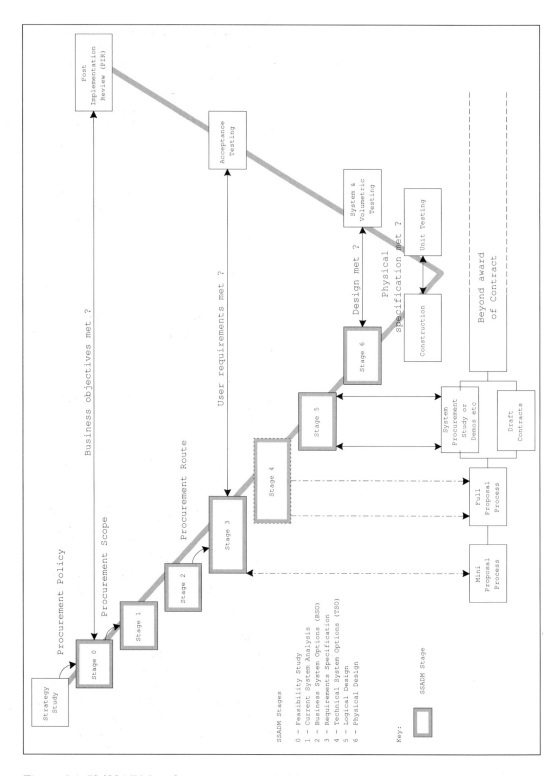

Figure 3.1: IS/SSADM and procurement activities

Formal input to the procurement process from the potential suppliers may occur as early as mid-way through Stage 3 Definition of Requirements. This formal input takes the form of a Mini-Proposal prepared by the supplier in response to a Statement of Requirement (SOR) for a procurement which is following the TAP Medium or High Complexity route. The SOR gives the supplier a precise statement of mandatory business and operational requirements.

The Mini-Proposals give the customer an outline of the suppliers' proposed solutions, background information on the suppliers and their products and valuable budgetary information on costs.

By following this Mini-Proposal process the customer and the suppliers gain certain other important benefits:

- the customer's understanding of how the requirement might be met is significantly enhanced at an early stage

- much of the costings information input to preparation of the Business Case comes directly from the suppliers

- suppliers who are not short-listed at the Mini-Proposal stage are spared the cost of preparing Full Proposals.

The second issue which the diagram highlights is that on procurement projects Stage 4 of SSADM is largely supplanted by the procurement activities. The selection of a Technical System Option (TSO) typically follows Stage 3. However, in a procurement project final TSO selection happens when the implementation contract is awarded.

In a procurement project some technical environmental considerations might need to be included in Stage 3. These could include IT standards and other design constraints determined by the IS strategy, or the number of terminals on the system, when this is directly related to user requirements. But even though taking decisions on these matters may be referred to as 'TSO decisions'

SSADM and Information Systems Procurement

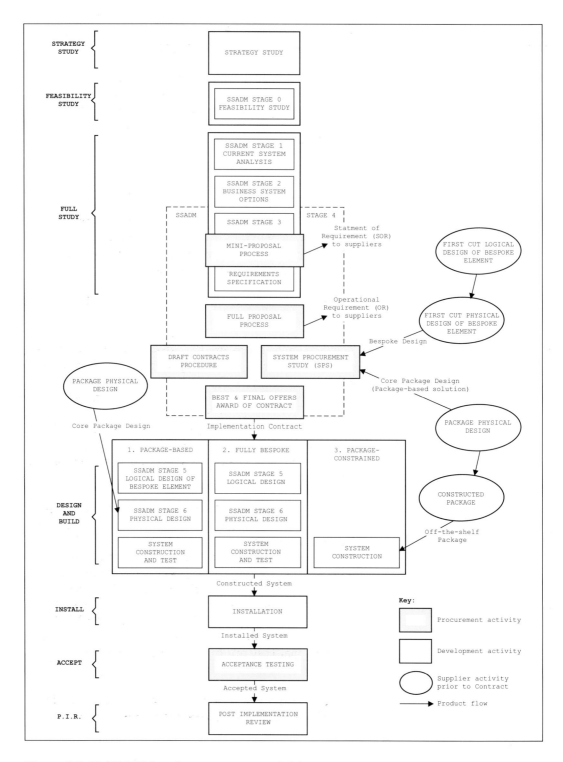

Figure 3.2: IS/SSADM and procurement activities

by some, a formal SSADM Stage 4 is not normally appropriate on a procurement project.

Figure 3.2 provides a more detailed view of SSADM and procurement activities in the development life cycle. However, this still represents a simplification. For example, in particular procurements there may be no Feasibility Study; and no attempt has been made on the diagram to show the maintenance loop.

Where Figure 3.1 showed the development activities as a 'V' and showed the procurement activities separately below the 'V', in Figure 3.2 the development and procurement activities are mingled together, and the diagram presents the activities in the chronological order in which they will normally happen.

Some of the elements of Figure 3.2 require a little explanation.

The terms which procurement specialists use to describe the development activities appear down the left-hand edge of the diagram. The curly brackets relate these names to boxes representing the development activities. All but one of the SSADM stages appear as boxes within the appropriate development activity box. Because it is supplanted by the procurement activities, Stage 4 Technical System Options is shown only as a perforated box.

Although Stage 4 and the related procurement processes start during, or after, Stage 3, customers need to consider for themselves whether certain high-level technical environmental issues, such as standards and strategies, may act as constraints on suppliers. Such high-level technical decisions may derive from the IS strategy, or be taken during the Feasibility Study, or at the time of Business System Options (BSO) selection in SSADM Stage 2, or as a mixture of these. However, Figure 3.2 assumes that they are finalized at the time of BSO selection.

There are instances where both procurement activities and SSADM Stage 4 are necessary, eg if the hardware is being procured, but the software is to be developed in-house, or vice-versa.

The procurement activities appear on the figure as a set of six shaded boxes:

- Mini-Proposal Process
- Full Proposal Process
- Draft Contracts Procedures
- System Procurement Study
- Best and Final Offers/Award of Contract
- Acceptance Testing.

The figure shows how the majority of the procurement processes happen between completion of the Requirements Specification and commencement of design and build of the new system. The Mini-Proposal process is an exception. The box which represents the Mini-Proposal process overlays the box which represents Stage 3. This is intended to highlight that the Mini-Proposal process happens during SSADM Stage 3. More details about when the SOR may be prepared are given in Section 3.3.

One of the main objectives of this volume is to give guidance on how the customer can reduce the time and cost of the procurement activities referred to as the Mini-Proposal Process and the Full Proposal Process, by ensuring that the SSADM products of the Requirements Specification can be 'pasted' into, respectively, the SOR and the OR. This is discussed in detail in Chapter 9.

Figure 3.2 shows the major procurement activities to be quite distinct from the Strategy Study. However, in practice, it is during the Strategy Study that the ground is prepared for the procurement process. For example, the relevant EC Directives are identified, and the procurement policy is agreed. It may also be during the Strategy Study that the ground is prepared for SSADM too. For example, the Strategy Study may identify

SSADM as one of the standards to be followed on the follow-up Feasibility and Full Studies.

Figure 3.2 also shows that the Feasibility Study (SSADM Stage 0) and early parts of the Full Study (SSADM Stages 1 and 2 and a little of Stage 3) are all undertaken before commencement of the first major procurement activity, the production of the SOR as part of the Mini-Proposal process. However, this is a simplification. Procurement considerations are likely to have an effect upon the project before production of the SOR. For example, during selection from among the Feasibility Options, consideration is given to the implications of procuring a solution. The customer may also continue to take procurement into account during SSADM Stages 1 – 3. For example, by making an initial evaluation of application packages, and by preparing the SSADM products at a level of detail appropriate to the procurement.

3.3 Producing the SOR and OR

This section describes in more depth the precise relationships between SSADM and the procurement processes of the production of the SOR and the OR. It also expands upon the variations in these relationships that may be needed in specific circumstances.

Figure 3.2 shows that the SOR is typically produced during SSADM Stage 3 and is part of the Mini-Proposal process. The SOR incorporates certain SSADM products, which are listed in Chapter 6. These include enough of the Required System Logical Data Model (LDM) and Data Flow Model (DFM) to enable potential suppliers to scope the system to be procured. Production of the SOR may, therefore, begin when SSADM Steps 310 Define Required System Processing and Step 320 Develop Required Data Model have been completed.

Figure 3.2 also shows that the OR is prepared at the end of SSADM Stage 3, that is, when the Requirements Specification is complete. This, however, is a simplification. For some procurements it may be appropriate to include in the OR some of the products of Stage 5. For example, if the requirement is for a bespoke system, for which a System Procurement Study (SPS) is not warranted, there may nevertheless be parts of the requirement which are complex and for which the detail of SSADM Stage 5 products would be helpful to the supplier.

For others it may be appropriate for the customer to carry out Stage 5 – as a means of fully understanding and 'teasing-out' the requirement – without including the Stage 5 products in the OR. Chapter 6 describes the SSADM products which may be incorporated into the SOR and the OR. For each an indication is given of its value, and from this readers may draw their own conclusions about the circumstances under which the SSADM product may or may not be included in the procurement product.

With regard to the design and build activities, the figure distinguishes between a package-based solution, a fully bespoke solution and a package-constrained solution.

The ISE Library volume: *SSADM and Application Packages* defines a 'package-based' solution as one in which the core user requirements are met by an application package, but other parts of the user requirements require bespoke additions.

In such a package-based solution SSADM Stage 5, if used at all, need only consider the bespoke element(s) of the required system. That is to say, for the required functionality which the package already provides, there is no need to specify the processing to the level of detail of, for example, an Update Process Model. Any products of Stage 5 are input to physical design in Stage 6, together with the design of the core of the application package. Thus, for the supplier of the package-based solution, a main objective of physical design is to integrate the physical design of the bespoke processes with the package's physical design.

In the case of a fully bespoke solution, SSADM Stages 5 and 6 may be relevant to the SOR and OR. The extent to which Stage 5 is relevant may depend upon the development environment – Chapters 6 and 7 discuss the use of Stage 5 in detail.

In the case of a package-constrained solution, neither SSADM Stage 5 Logical Design nor Stage 6 Physical Design are relevant. As defined in the ISE Library volume: *SSADM and Application Packages*, in a package-constrained solution the user requirement is constrained so that the 'off-the-shelf' version of the package meets the requirement. The 'development' activity required is, therefore, likely to be limited to integrating the package with the customer's existing systems.

Both Figure 3.1 and Figure 3.2 are intended to highlight the role of SSADM Stages 0 – 3 as a means of specifying the requirement in a procurement project. However, this is not to say that SSADM is appropriate to all types of procurement. The most common types of procurement are listed in Chapter 4, and an indication is given for each of the usefulness of SSADM.

3.4 Management decision points

A procurement project does not affect the timing of decision points, but it does have an impact upon the scope of the issues to be considered.

- Decisions about the scope of the development and the business objectives to be addressed are taken either at the Strategy Study or at the Feasibility Study stage. These should include procurement policy decisions at the Strategy Study stage and procurement scope decisions at the Feasibility Study stage.

- Agreeing the operation of the current system, problems in the operation of the current system and the functional and non-functional requirements for the new system occurs at the end of Stage 1 Investigation of Current Environment, and is unlikely to be affected by the procurement.

- Choosing the scope of the information system requirements at a cost acceptable to the customer occurs in Stage 2 Business System Options. In addition, the procurement strategy and the procurement route must be agreed at the time of BSO selection.

- Agreeing the information to be managed by the new system, the business activity patterns to be represented and the information management and access facilities to be made available to end-users and other systems occurs at the time of signing-off the Requirements Specification, which is either at the end of Stage 3 or at the end of Stage 5. Chapter 6 discusses the role of Stage 5 in the Full Study; the requirement issues agreed at this point must be supplemented by further decisions in connection with the procurement, and in particular the procurement and contractual requirements to be included in the OR.

- Selecting the hardware and software platform for running the new system is greatly affected in a procurement project. Instead of TSO selection in SSADM Stage 4, the customer undergoes a series of transactions with potential suppliers, which may begin formally with the Mini-Proposal process, and which ends for the purposes of this volume with the award of the implementation contract to the selected supplier. Alternatively, hardware and software platforms might be called-off from existing Framework Agreements.

- Agreement to the computer system design, which implements the Requirements Specification, occurs at the end of Stage 6 Physical Design. This agreement is, however, affected by the contractual agreement between the customer and the supplier. Under the contract, Stage 6 will be the responsibility of the supplier, therefore the only basis on which the customer might review Stage 6 products is to confirm the supplier's understanding and interpretation of the customer's specified requirements. The review of the design is not an acceptance of the solution and, therefore, the supplier's design, which must only take place on successful completion of acceptance trials. These must not be pre-empted by earlier customer agreement to design documentation.

3.5 Procuring high complexity systems

High complexity and, therefore, high risk projects, will follow the TAP High Complexity procurement route. The essential difference between this procurement route and lower complexity routes is the inclusion of a System Procurement Study (SPS). The purpose of the SPS is to enable the short-listed suppliers to gain greater understanding of the requirements and to develop their design of the solution to a stage that permits them to produce a fixed price quotation. It also enables the customer to gain a greater understanding of, and greater confidence in, the solutions being proposed by the suppliers. SSADM products and techniques are likely to be very relevant to the SPS.

Figure 3.2 shows the activities connected with the SPS. It includes, as ovals, the design activities of suppliers which happen before the award of contract. The products of these activities are input to the procurement activities. For example, if the supplier is required under the SPS contract to prepare a first cut logical design using Stage 5, then, as the diagram shows, the supplier produces a bespoke logical design for part or all of the solution which is converted into a first cut physical design and then input to the SPS activity. Chapter 7 discusses in detail the circumstances under which Stage 5 activities might be undertaken by SPS suppliers. The final logical design is completed after award of contract.

The SPS activity is also likely to involve both first cut physical design and a certain amount of construction for demonstration purposes. Demonstrations within the SPS may also include bespoke and package elements of the application software.

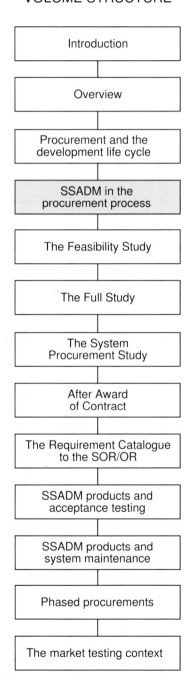

4 SSADM in the procurement process

4.1 The role of SSADM in the procurement process

As an analysis and design method, SSADM has an obvious role in the procurement of systems that include a significant amount of bespoke development; thus SSADM can make a direct contribution to the procurement process. In addition, SSADM may be used in a number of other types of system procurements. It may also be of benefit to the procurement of other types of support services.

SSADM is intended primarily to prepare for the construction of information systems. In procurement terms this means that SSADM is primarily applicable to those projects which result in bespoke development. Its most powerful techniques focus on producing the correct specification of an organization's requirements for an application database and software. It considers in great detail the data requirements and the functional requirements; and it encourages users and analysts to specify the service level requirements, though mainly those directly related to the functionality of the system. The specification techniques in SSADM dovetail with a set of follow-up techniques covering the design of the database and software.

For the purposes of defining the types of procurement to which SSADM is applicable it is possible to place the most common procurement types on a sliding scale. At one end of the scale is the fully bespoke development, for which thorough analysis and design are required, and to which SSADM definitely applies. At the other end of the scale are those procurement types with no element of analysis and design, and to which SSADM definitely does not apply; in between is a set of procurement types with varying amounts of analysis and design, for which the applicability of SSADM will correspondingly vary (see section 4.3).

An additional complexity occurs where a procurement is, in practice, a mixture of procurement types. The customer's requirement may be met partly by procuring bespoke software, and partly by procuring an application package. The customer may wish to procure not only software but also hardware on which to run the operational system. In such cases SSADM may be relevant to some parts of the procurement, but not to all of the parts.

The following sections describe the extent to which SSADM is applicable to the most common types of procurement. Section 4.3 refers to the procurement of systems. Other terms used to describe this type of procurement are 'supply' procurements and 'goods' procurements. Section 4.4 refers to the procurement of services. Each is presented in descending order of the usefulness of SSADM, and must be used in a 'mix and match' way on projects involving multiple procurement types.

In the case of procurements by public sector organizations where the value of what is being procured is greater than the EC/GATT threshold, either the EC Supplies Directive or the EC Services Directive will apply. The classifications in the following sections are not intended to indicate which Directive will apply to a procurement type. Guidance should be sought on this matter on a procurement-by-procurement basis.

4.2 Procurement and the 3-schema specification architecture

As regards the procurement of bespoke, package-based and package-constrained systems the 3-schema specification architecture is little changed. As detailed in Section 2.2, what is changed is mainly the division of the activities between the customer and the supplier.

The system development template of the 3-schema specification architecture presumes the seamless transition from the specification of the requirement into the design of the solution.

In a procurement project this seamless transition can be broken by the interface between customer and supplier. Especially so in a procurement under the EC Supplies Directive or EC Services Directive, where the employment of SSADM in design activities by the supplier cannot be specifically stipulated.

The effect on the 3-schema specification architecture of the procurement of an application package is described in the ISE Library volume: *SSADM and Application*

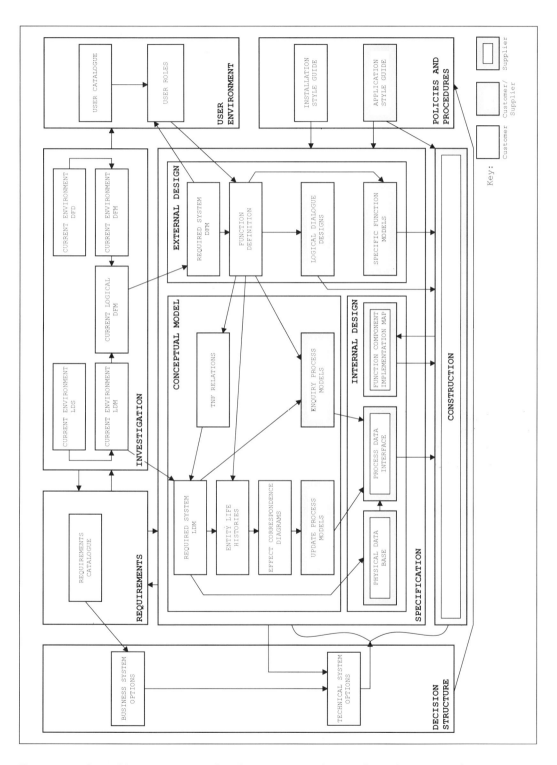

Figure 4.1: Core SSADM system development template and 3-schema specification architecture in a bespoke procurement project

Packages. Figure 4.1 illustrates the effect of the procurement of a bespoke system on the core SSADM 3-schema specification architecture template.

The default template is little changed in structure. However, in recognition of the importance of the Requirements Catalogue this is separated from the balance of the investigation processes, and the iterative completion of the catalogue is indicated by the two-way arrows between it and the investigation and specification processes. The change in the function of the Technical System Options stage is also illustrated. As indicated in Chapter 2, this stage is effectively replaced by the procurement process, which will detail and drive all of the supplier activities. This is indicated by the bracketing of these activities on the figure.

The only other changes are in the use of different outlines to indicate which processes are the responsibility of the customer, which the responsibility of the supplier and which can be completed by either of them. This is discussed more fully in Chapter 6.

4.3 SSADM in systems procurements

In order to help readers to understand where SSADM can deliver real benefit to the procurement process, this section categorizes systems procurements and describes the contribution that SSADM can make to each category. The categories are shown in descending order of the potential contribution that SSADM can provide.

Fully bespoke solution
In this type of procurement the customer wishes to buy an IS to support business functions which few other organizations carry out and for which, therefore, no application package is available. Full analysis and design will be required. Therefore, SSADM is fully applicable.

Package-based solution
In this type of procurement SSADM Stages 0 – 3 are applicable as a means of specifying the requirement. Guidance on the use of SSADM in this type of project is given in the ISE Library volume: *SSADM and Application Packages*. The outputs of Stages 0 – 3 also have the potential to be used in the evaluation of packages in the procurement process. For example, how closely the package fits the requirement may in part be assessed by comparing the Logical Data Model with the packages's

physical data model. The SSADM products may also be helpful in the paramatization of the package procured.

The extent to which the SSADM design stages will be useful is less clear cut. In the case of packages which include straightforward functionality, such as record keeping systems, the design stages may be omitted if:

- the extensions/modifications to the package can be described in sufficient detail using SSADM Stage 3, specification products only

- the customer is not procuring the source code: that is, the supplier is responsible for maintenance of the software.

In the case of packages which offer core functionality for a specific vertical market (such as banking and underwriting) and which are coupled with powerful application generator facilities, the SSADM design stages are more likely to be helpful in extending or modifying the package.

In the operational phase, customer input to the maintenance of the software is made easier, if the supplier has prepared an SSADM design. Indeed, in some cases the customer may procure the design and source code, and maintain the system in-house. (Alternatively, the customer might procure the design and source code, and enter into a contract with a third party to maintain the system.)

Note, that if the supplier does not maintain an SSADM design, the customer should establish the standard to be applied in assessing the quality of the system and its enhancements. Whether by means of an SSADM design or not, the customer needs a general understanding of the design of the system, in order to avoid over-dependence on the supplier's knowledge of the system.

Package-constrained solution	A package-constrained solution is one which is constrained to only those functions that can be provided by application packages.

Some or all of SSADM Stages 0 – 3 may be applicable as a means of specifying the requirement, depending upon how early in the development life cycle the customer identifies that a package-constrained solution is acceptable. The outputs of Stages 0 – 3 also have the potential to be used in the evaluation of packages in the procurement process, and in the paramatization of the package procured. The SSADM design stages are unlikely to be applicable.

Guidance on the use of SSADM in a package-constrained solution is given in the ISE Library volume: *SSADM and Application Packages*.

Standard infrastructure packages	In this type of procurement the customer is typically seeking to procure Office Automation packages, such as word processing and spreadsheet packages. This type of procurement is not concerned with the design of software and the design of a database. Therefore, SSADM is largely inapplicable (although an SSADM Requirements Catalogue may be prepared, and other SSADM techniques may be used to describe the flow of data between the different components of the infrastructure).
Hardware procurements	This type of procurement involves neither analysis nor design. SSADM is, therefore, not strictly applicable, although it may be relevant to application sizing.
Communications software	In this type of procurement the customer is procuring, for example, netware to control communications across a Local Area Network. SSADM is not applicable.
PBX	In this type of procurement the customer is procuring, for example, a telephone switchboard. Despite the increasing sophistication of modern PBX systems and the facilities for handling data (such as stored messages), the procurement involves no analysis or design in SSADM terms. SSADM is, therefore, not applicable.
Development software	In this type of procurement the customer is procuring, for example, a 4GL or RDBMS. SSADM is not directly applicable, but the ability of the software to produce SSADM products may be of interest. Guidance is given in the SSADM Tools Conformance Scheme.

4.4 SSADM in services procurements

Readers are reminded that this volume deals with the use of SSADM in the procurement process. It is not directly concerned with the procurement of SSADM services. The potential benefit and use of SSADM is limited in services procurements other than those involving bespoke software development. However, there are categories of services procurements where SSADM can be of benefit. Service categories are shown in descending order of the potential contribution that SSADM can provide.

Software-based service

In this type of procurement the customer procures software developed by the supplier as a result of the contract. In some cases the supplier may have to develop all of the software needed to meet the requirement – in which case there is no difference between this type of procurement and 'fully bespoke solutions' in Section 4.3. In other cases the supplier may have to develop software to enhance an application package, or the supplier may have to modify the package.

To the extent that the software production includes the design of the solution, SSADM is as applicable to this type of procurement as it is to the fully bespoke or package-based procurements referred to in Section 4.3.

Maintenance

Maintenance procurements are where the customer requires the supplier to maintain one or more systems which may, or may not, have been built from an SSADM design. SSADM as a method does not apply to such procurements. However, it is important to include available SSADM documentation as part of the procurement documentation to assist potential suppliers more fully understand the system, or systems.

Examples of this type of procurement are:

- third party maintenance: that is, where the customer procures services for the maintenance of hardware and/or network(s) from someone other than the original supplier

- software maintenance.

If one or more of the systems to be maintained has been built from an SSADM design and the customer wishes to use that design as a means of input to the supplier's maintenance activities, then the customer and supplier must agree which SSADM products will be maintained under the contract. In this instance SSADM is not applicable to the procurement itself (it is used by neither the customer nor the supplier during the procurement project). **However SSADM is applicable to the operation of the maintenance contract which results from the procurement.**

Systems integration

In this type of procurement the customer requires the supplier to integrate two or more existing systems. These may or may not have been developed using SSADM. For those systems for which SSADM documentation exists, enough of the documentation must be included in the procurement products to enable the suppliers to scope the integration task. The physical data designs and system sizing documentation will be particularly helpful to the suppliers.

Outsourced IS/IT service provision

This term applies to a wide range of services, which may include development services. The term may be used to refer to services procured as a result of market testing, and it may be used to refer to the following 'traditional' types of service:

- facilities management 'outsourcing' or 'contracting out'

- bureau services

- managed network: that is, where the customer procures a network monitored and controlled by the network provider into which raw data is fed and out of which raw data comes

- Value Added Data services: that is, where the customer procures a network monitored and controlled by the network provider, into which raw data is fed, but out of which other data than the raw data comes

- disaster recovery, for example, where the customer procures bureau facilities on which to run systems in the event of disaster on the normal host environment.

SSADM is applicable to the operation of the contract in ensuring that the service providers, where necessary, are able to maintain existing systems or implement new systems using SSADM.

Design recovery
In this type of procurement the customer requires the supplier to define the design of a system, which is either undocumented or documented at only the physical level. Where the supplier is asked to retro-fit SSADM design products, the customer and supplier must agree which SSADM products will be developed under the contract. Design recovery procurement is another instance in which SSADM is applicable to the operation of the contract which results from the procurement, rather than the procurement itself.

Consultancy services
This type of procurement falls outside the scope of SSADM, although it may be SSADM skills which are being bought in.

4.5 Benefits from the use of SSADM

Apart from the well understood benefits which a structured approach brings to any project, procurements where the use of SSADM is appropriate can directly benefit from its use in a variety of ways. This section indicates the main areas of benefit.

Full description of the requirement
SSADM takes three views of the required system, its data, its processing and the effect of the processing on the data over time. The cross-validation between these three views often reveals both additional details and additional requirements which are not apparent from any one view taken in isolation.

Precise description of the requirement
Many of the SSADM products are based on Jackson diagrams, which introduce a degree of precision which is difficult to match in narrative alone. Specifications based on diagrams rather than narrative are particularly important given that suppliers will not necessarily be based in the UK. Where requirements are expressed in

narrative, SSADM encourages the use of measurable terms wherever possible.

Common terminology	As the most widely used analysis and design method in the UK, SSADM provides a *lingua franca* by which suppliers can understand customers' requirements, and by which customers can understand suppliers' designs where these have been developed using SSADM. However, it may not be possible to mandate the use of SSADM for systems procured under EC regulations.
Service level requirements	Beginning in the Feasibility Study SSADM encourages the specification of the non-functional and service level requirements. (An example of a non-functional requirement might be, 'the operator must receive a screen response within 2 seconds'.)
Benefits linked to requirements	One of the most important products of SSADM is the SSADM Requirements Catalogue, which acts as a central repository of all the requirements. The Requirements Catalogue is opened in Stage 0 Feasibility, and is continuously developed throughout the Full Study. Analysts are encouraged to describe – for each entry in the Requirements Catalogue – the benefits which that requirement will bring to the customer. This can help to ensure that the procurement focuses on those requirements that will most benefit the organization.
Prioritization of requirements	SSADM encourages users to prioritize the requirements, in order to make the essential, core requirements easily distinguishable from the 'wish list' ones. This prioritization can commence as early as the Feasibility Study, when the Requirements Catalogue is opened. Chapter 9 describes the use of the Requirements Catalogue on procurement projects, including the importance of using 'mandatory' and 'desirable' as the priority levels for procurement purposes.
Use of SSADM products in products sent to suppliers	When an SSADM Requirements Catalogue is available, the amount of time needed to draft the SOR/OR can be much reduced. Chapter 9 provides detailed guidance on how to draft the Requirements Catalogue in a way which makes its entries suitable for 'pasting' into the system requirement sections of the SOR and OR.

	Other SSADM products can be suitable for inclusion in the SOR and OR as annexes – Chapter 6 gives guidance on this matter. Some are suitable as a means of amplifying the requirements; others are not suited to this purpose, but are useful as a means of providing suppliers with background information to help them understand the requirements.
Ease of maintenance	All of the SSADM products assist in the transformation of the requirements expressed in the Requirements Catalogue into implemented IT support for the requirements in an operational information system. If the customer's requirement is specified using SSADM and if the supplier designs the solution using SSADM, then the maintenance effort should be significantly reduced by the ease with which the design can be understood.
	The ease of understanding the SSADM design and the SSADM specification from which it has been derived, is in part the result of the requirements traceability paths which the SSADM products support. Chapter 10 describes the SSADM requirements traceability paths.
Hooks for other methods	SSADM has sufficient hooks to enable other development methods to be used on the same project. An example of where this might be an advantage is where a tailored package is to be procured, but the bespoke tailoring cannot be done using SSADM, but the package is to fit into an internally developed SSADM-based system.
	SSADM's physical design products can be used to refer to non-SSADM elements of a design. For example, the Function Component Implementation Map can be used to document which functions have been implemented by the package and how they fit in with the SSADM-based functions. Also, a Process Data Interface can indicate how accesses to entities in the Logical Data Model are implemented as accesses to a non-SSADM physical data model.
Acceptance criteria	Acceptance criteria for the installed system are based on the requirements of the system itself as stated in the OR. If these OR requirements are chiefly developed from the Requirements Catalogue, the majority of acceptance

criteria can be documented in the SSADM specification products. However, there will be other acceptance criteria which cannot be documented in SSADM products, for example, quality standards for manuals.

Chapter 9 of this volume gives guidance on how to express requirements so that they are measurable. Chapter 10 gives guidance on how to build up a requirements traceability path from the entries in the Requirements Catalogue. If this guidance is followed, then SSADM products can be used during the Full Study to document the acceptance criteria for the data, functional and non-functional requirements.

4.6 Limitations in the use of SSADM

Procurement is a complex process that requires a variety of skills. Whilst SSADM can deliver many benefits to the procurement process, it should not be considered a panacea. There are very specific limitations to the benefits of its use. This section describes the most frequently held expectations of SSADM which are often without foundation. In addition, the use of SSADM introduces cultural issues which must be managed.

The right requirement

The use of SSADM must be supplemented with knowledge of the business domain and good analytical skills.

Even if the customer has the right requirement in mind, the use of SSADM does not ensure that the requirement will be well specified. For example, despite guidance from the method to the contrary, many customers continue to express their requirements in terms of possible solutions, and to a level of detail which unnecessarily constrains suppliers' solutions.

SSADM contains many diagrammatic products with which the processing and data requirements can be expressed very precisely. There is no guarantee in any method that its products specify the requirement in sufficient detail, and free from ambiguity, to avoid contract variation claims by the supplier through misinterpretation of imprecise requirements.

It is important to ensure a common understanding is developed between customer and supplier during the

Technical Discussions and Demonstration phase of TAP. On procurements following the TAP High Complexity route, the System Procurement Study gives the supplier an opportunity to gain a further level of understanding of the requirement.

Handling suppliers

Nothing in SSADM helps the customer to assess the suppliers' capabilities. Familiarity with the market is required on the part of the customer.

SSADM does not cover all of the issues which need to be considered at the time of drawing up a Service Level Agreement, for example, the turn-round times for fixes to bugs in the software.

Culture gaps

SSADM, or indeed, any other method, brings with it an additional layer of terminology. This may introduce difficuties for procurement specialists and suppliers who are not familiar with that method.

In order that the introduction of such new terminology does not form a barrier to communication, this volume seeks to provide a greater understanding of procurement issues to the management and SSADM communities and to bridge any existing culture gaps between the two communities.

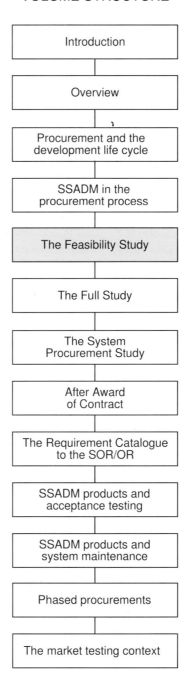

5 The Feasibility Study

SSADM strongly recommends that a Feasibility Study is carried out before a Full Study, unless the proposed system is low risk. During the Feasibility Study the customer has an opportunity to identify procurement possibilities, and, if any exist, to identify the constraints which would affect the procurements, such as the relevant EC/GATT regulations, and the standards imposed by the customer's own organization.

Procurement of a solution, especially one based wholly or to a great extent on the (existing and proven) design of an application package, can help to reduce risks compared with in-house development. Procurement, especially a phased procurement, should always be considered among the Feasibility Options wherever there is a need to reduce risks, such as:

- technical risks

- project management risks

- cost risks

- evolving requirements.

The same SSADM products are produced during SSADM Stage 0 Feasibility whether procurement is considered or not. Where procurement is considered, the SSADM products may need to be supported by additional, non-SSADM products, the development of which must be planned and controlled.

At the time of preparing and selecting from Feasibility Options which include procurement possibilities, attention should be given to which SSADM stages would be necessary in the Full Study and how these would dovetail with the preparation of the products of the procurement process, and the selection of an appropriate procurement strategy.

5.1 Role of the Feasibility Study

A Feasibility Study is a short assessment of a proposed information system to determine whether the system can effectively meet the specified business requirements of the organization, and whether a business case exists for developing or procuring such a system.

The Feasibility Study should normally begin the process of determining whether any or all of the products and services required to develop the information system could be procured. If procurement possibilities are identified, then the impact of the procurement(s) should be described in the Feasibility Options which are presented for selection.

During the Feasibility Study the customer should, therefore, take the opportunity to initiate contact with potential suppliers to explore the procurement possibilities. However, contact with suppliers at this stage does not form part of a formal tendering process.

A Feasibility Study may either be initiated by an IS Strategy Study, or it may be initiated by the identification of an IS requirement from within a part of the business.

Where procurement possibilities have been identified and the Feasibility Study follows an IS Strategy Study, the Feasibility Study must take into account the products input to it from the IS Strategy Study, which may place constraints upon how a procurement would have to be conducted. The products from the IS Strategy Study which may need to be considered, and the procurement constraints which they may introduce, are described in the following section.

Where procurement possibilities have been identified but the Feasibility Study is being conducted by an organization which has not adopted an IS Strategy Study, the Feasibility Study must consider the same range of issues, and seek to reach agreement on what constraints will apply to a subsequent procurement.

5.2 Procurement issues for the study

Feasibility Study projects which follow an IS Strategy Study may be informed by IS Strategy documents which raise all or most of the procurement issues, and which are described in brief in this section. In such cases the Feasibility Study Initiation Document should either include relevant passages from these documents, or it should cross-refer to them.

For Feasibility Study projects which do not follow an IS Strategy the preparation of the Feasibility Study Initiation Document (FSID) requires that additional resources be expended. The FSID should include all of the customer's relevant standards, policies and constraints. Where none exist at the organization level, these matters must be drafted into the initiation document and management agreement to them secured.

Strategy level plans

The customer may have drawn up strategy level plans affecting its information systems. These plans are likely to include timescales which place constraints on projects and procurements, eg 'the new budget accounting systems must be available before the EC regulations change in January 1995'. These constraints must be taken into account when outline plans are developed for the Feasibility Options.

The Feasibility Study must also consider to what extent the organization is seeking to protect its investment in existing software developed in-house, and how committed the organization is to retaining resources with which to maintain IT systems.

Technical policies

At the strategic level, the customer may have adopted technical policies which refer to such matters as target development environments and target operational environments. For example, as a result of the IS Strategy the organization may run a competition which results in the selection of a 4GL toolset for use on certain types of development projects.

Alternatively, the IS Strategy may have identified specific IT facilities that should be used to meet certain types of requirement, for example, that an existing mainframe facility should be used to support operational systems of a certain size. These policies will constrain the nature of the candidate projects and the extent of procurement possibilities which can be presented to management in the Feasibility Options.

Procurement policy

At a strategic level an organization may set policies which impact on the way individual procurements are to be undertaken. For example, the IS Strategy may require that all procurements must be phased if the total software and hardware purchase cost is likely to exceed, say, £2m. Or the IS Strategy Study may have identified that the solution to a particular business requirement should be purchased, because, for example, it is known that insufficient in-house skills exist for an in-house development.

EC/GATT regulations

Where a Feasibility Option describes a candidate procurement which is likely to exceed the EC/GATT limits, the description of the option must take the regulations into account. In particular, the description of the Feasibility Option should not assume the use of SSADM, since under both the Supplies Directive and the Services Directive it is not possible for customers to specify that suppliers use SSADM as a mandatory requirement, although it can be specified as a desirable requirement. In addition, the outline plan associated with the Feasibility Option should take into account the time constraints which will be imposed on the project timescales by the procurement method. More detail on these constraints are included in *A Guide to Procurement within the Total Acquisition Process* written by CCTA.

5.3 The SSADM products

The same SSADM products are prepared during Stage 0 Feasibility, whether procurement possibilities are being considered or not. However, where procurement is likely to form part of the project selected for the Full Study, some adjustment should be made to the format of entries in the SSADM Requirements Catalogue.

With one exception, the SSADM techniques and products of Stage 0 Feasibility are unaffected by the consideration of procurement possibilities. The exception is the SSADM Requirements Catalogue.

Where it is possible that a procurement Full Study project using SSADM will be selected (from among the

Feasibility Options), entries in the Requirements Catalogue should be drafted from the outset with their appropriateness for inclusion in the procurement products in mind. The formatting and expression of entries in the Requirements Catalogue in readiness for inclusion in the SOR and OR is described in Chapter 9.

5.4 Supporting non-SSADM products

When the Feasibility Options are to include candidate procurement projects, it is necessary for the activities of SSADM Stage 0 Feasibility to be supplemented with additional activities. These will be aimed at gathering information from suppliers about the procurement possibilities. However, it is usually the case that the amount of time available for the Feasibility Study is restricted. Therefore, the additional activities must be carefully planned, monitored and controlled.

The products of the additional, procurement-oriented activities are described in this section. If these products are to be developed the Project Plans should include appropriate Product Descriptions, each with quality criteria and a definition of the amount of detail to be documented.

Information gathering

Where the technical knowledge available to the Feasibility Study team is limited – from within the team itself and from other parts of the customer organization – it is essential to supplement the knowledge with input from suppliers. Such input may be requested either informally or formally.

- **Informal request for information from suppliers**
  This can be a simple and quick method of obtaining information, but it is unlikely to yield comprehensive and fully reliable information, since it is not usually practical to give the suppliers a comprehensive description of the requirements. In practice, the issuing of informal requests for information and receipt of responses to them happens iteratively and takes many different forms, from written exchanges to telephone contact, and from submission of brochures to demonstrations of available packages. For procurements that are likely to be above the EC/GATT thresholds, it should be

made clear that this information gathering is not part of a procurement process.

- **Formal Request For Information (RFI)**
  This involves advertising the RFI in the European Journal. The information so collected can be more comprehensive than the informal method, but it takes longer to collect. If a procurement follows the RFI, that procurement must be separately advertised in the European Journal.

These informal and formal methods of obtaining information are not mutually exclusive, and combinations of methods can be used.

Requests for information may provide the Feasibility Study team with technical guidance on a wide range of issues. For example:

- what application packages are available that might meet the business requirement

- what technical or business constraints these packages may impose

- what hardware and software is available to support the technical options being considered

- what products are available to support distributed systems.

The Feasibility Study team's increased technical knowledge gained from early contact with suppliers might result in significant changes to the Feasibility Options. For example, where a fully bespoke development had been assumed, it may be possible to present as a Feasibility Option the procurement of a package-based solution.

Requests for information not only broaden the Feasibility Study team's technical knowledge but also, equally importantly, give the team information about the suppliers' technical strengths, size, track record and the like, which helps to inform the risk assessment portion of Feasibility Options.

	Supplier assessment	As a result of knowledge about suppliers gained during the Feasibility Study, whether directly from the suppliers themselves or indirectly from third parties, the team may prepare an informal supplier assessment which reviews the financial and/or technical acceptability of suppliers. Such a product is useful if a Feasibility Option includes the procurement of an application package or the procurement of development services. However, the EC/GATT directives do not allow such informal assessments to restrict the subsequent competition.
		If either the EC Supplies Directive or the EC Services Directive applies, and if TAP procedures are followed, formal supplier assessment must form part of the procurement procedures.
	Application package assessment	As a result of knowledge gained about application packages considered during the Feasibility Study, the team may prepare assessments of how closely particular packages meet the business and technical requirements. The amount of time available for the evaluation of application packages during a Feasibility Study is often very restricted, so the assessments are likely to be expressed in comparative terms only.
		Further guidance on the evaluation of application packages is given in CCTA's Appraisal & Evaluation Library volume: *Overview & Procedures*.
5.5	Documenting the options	When preparing Feasibility Options for selection it is important to distinguish which represent procurement options. A Feasibility Option may be to procure all of the solution to the requirement, or to procure a solution to only part of the requirement, or one which does not include any procurement. However, assumptions made in the Feasibility Study about the extent to which the solution will be procured may be subsequently overturned as a result of further analysis during the Full Study.
		Feasibility Options which describe candidate procurement projects must address a different set of risks from those which describe in-house development projects. In particular, there will be risks associated with the implementation contract, which must define clearly

the customer's and the chosen supplier's responsibilities. The amount of time and the expertise needed to draw up such a contract should not be underestimated. In addition, there may be risks associated with the supplier's financial stability.

The content and layout of Feasibility Options may follow the customer's own documentation standards, or, if no local standard applies, the Feasibility Options may be documented along the lines suggested in the SSADM manuals (page F-FS-19 of the manuals refers). Irrespective of their layout the Feasibility Options should cover at least the topics referred to in this section.

Scope of the information system

A description of the scope of the proposed information system with particular reference to its clerical and IT aspects. This section should be supported by subsets of the Data Flow Model and the Logical Data Model produced as part of the Outline Required Environment Description. The reasons for the proposed scope should be clearly stated, especially if the proposed scope is set as a result of procurement considerations.

The requirements and performance measures addressed by the Feasibility Option should be described. It should be possible to take them from the Requirements Catalogue and 'paste' them into the Feasibility Option documentation.

An essential part of expressing the scope of an information system is defining who in the customer organization would be affected by the proposed information system. To highlight this issue, the Feasibility Option should be supported by appropriate entries from the SSADM User Catalogue. Clarity on this aspect of the scope is particularly important in circumstances where a package-constrained solution might be acceptable. It is the managers in the business areas who should take the decision whether or not the tasks and working practices of their staff are sufficiently flexible to allow for the constraints imposed by any chosen package.

Hardware and software configuration overview	A textual description of the nature of the hardware, software and communications components of the proposed system, together with the physical location of the components. An indication of the system sizing should be provided. Depending on the information available, the system sizing information may be comparative only at this stage.

The description should make it clear which of the components would have to be procured. It should also indicate the likely extent to which the procured components would have to be integrated by the supplier. For example, it may be that a Feasibility Option is based on the procurement of an application package to run on the customer's current platform and network. But, if it is known that none of the application packages available has ever been installed on such a platform, then it is important to highlight the risks associated with the integration issue.

The description should also indicate the technical resources necessary to build and install the configuration. This may highlight procurement issues, such as the need for specialist (contractor) resources in the IS project team, or the dependence on supplier-developed enhancements to application packages.

It is important that the description of the configuration and sizing issues does not introduce a solutions bias into any procurement product. |
| Impact analysis | An outline assessment of the impact of the option on the customer organization. The focus of the impact analysis is normally on changes in the users' organization structure and working practices. However, where the Feasibility Option involves procurement, selection of the option might affect other parts of the customer organization. For example, the customer's IT services division may be required to re-allocate resources if it is proposed that an application package should replace a system currently maintained in-house. Alternatively, the customer's contract specialists may be required to exercise new skills if a type of procurement is proposed which is new to the customer. For example, if a turnkey |

Risks	solution is proposed and the customer has never procured a turnkey solution before.

A description of the major risks associated with the Feasibility Option in terms of business, technical, financial or cultural factors, together with an outline description of appropriate risk reduction and avoidance methods.

Options based on procurement will require the consideration of different types of risk to those associated with an in-house development. For example, there may be risks associated with the financial stability and technical competence of potential suppliers – see 'Supplier assessment' in Section 5.4.

There may also be risks associated with the drafting of a contract, especially for the provision of development services. A project to develop an information system is by nature rather like a research and development exercise – the final product is not known at the outset but evolves during the project. It may be attractive to the customer to hand over the risk of the 'R&D' to a supplier, but it is very difficult to draw up a contract which anticipates all of the circumstances which may be encountered.

For contracts under which a supplier agrees to undertake any or all of:

- requirements specification
- external design
- internal design

of an information system, it is very difficult to draw up a contract which makes it clear, under all circumstances, which activities are covered by the contract and where the division of responsibility lies between the customer and the supplier. The amount of time and the expertise needed to draw up such a contract should not be underestimated. |

As requirements specification is normally the responsibility of the customer, it is suggested that, if requirements specification is to be contracted out, it is done so under a separate contract. This will ensure that the very different responsibilities of requirements specification, and those of system design and implementation are clearly separated. However, this suggestion does not apply to the modification of an application package.

| Outline development plan | An outline timescale for implementation is essential, and if possible should be supported by an outline: |

- user implementation plan
- take-on plan
- data conversion plan
- training plan.

Where procurement is proposed the plans should indicate which activities will be the responsibility of the supplier(s). The plans should highlight instances where an activity to be undertaken by a supplier depends on completion of an activity by the customer, since additional charges or claims for damages under the contract may be incurred if the customer activity is not completed on time.

The outline development plans prepared for the candidate Full Study projects must take into account the different uses of SSADM that would be made on those projects. With regard to a project to develop the solution in-house, it should be possible to make reasonable assumptions about the use that will be made of SSADM, and, therefore, of the duration of the SSADM activities, especially if the organisation has developed its own 'templates' for the flexible use of SSADM.

Preparing outline development plans for candidate procurement projects may be more difficult, especially where the procurement is likely to be subject to EC/GATT regulations. Reasonable assumptions may be made about the duration of the (customer's own)

SSADM Full Study activities. Section 5.6 gives an overview of the circumstances which can significantly affect the appropriateness of SSADM Stage 1 Investigation of Current Environment and Stage 2 Business System Options.

However, it is more difficult to make estimates about the duration of the chosen supplier's development activities, since these may or may not be SSADM-based due to the restrictions of the EC/GATT regulations on the mandatory use of SSADM.

Another difficulty in the preparation of outline development plans for candidate procurement projects is knowing how much time to allow for the preparation of procurement products. The number of procurement products that will be required depends upon the chosen procurement strategy, 'big bang' or phased, and the procurement route, High, Medium or Low Complexity. Yet final decisions about which strategy to adopt and which route to follow are usually not taken until after the Full Study has begun, typically at the time of Business System Option selection.

At Feasibility Study stage only best estimates can be made. Advice should be sought from procurement specialists and all assumptions made in drawing up the development plan should be stated.

Investment appraisal

Many organizations have their own standards for the presentation of investment appraisals. Most, however, are based upon the CCTA standards described in the IS Guide B4: *Appraising Investments in IS*. Typically all investment appraisals show, on a year by year basis, the costs of a proposed system together with its associated benefits. At the Feasibility Study it is likely that only the broad order of the costs can be estimated, though the precision of the figures may be improved where estimates have been provided by suppliers in connection with formal or informal Requests For Information.

Since the return on the organization's investment identified by these figures is likely to be one of the most significant factors in the selection of a system to take forward into a Full Study project, the greatest possible effort should be made to include estimates of all of the costs and all of the benefits.

The investment appraisal created at this stage will be refined at least twice during a subsequent procurement project to provide a more precise and reliable appraisal:

- at the time of Business System Option selection, ie when the customer chooses an information system solution to some or all problems and requirements

- at the time of award of the implementation contract, when separate investment appraisals are drawn up for each supplier with whom draft contracts have been agreed. The costs at this point should reflect the 'best and final' offers of the suppliers over the full life of the system, and the benefits should be the anticipated benefits of the actual systems which the suppliers are committed to supplying under their Draft Contracts and associated schedules.

Depending on the project circumstances, there may be other decision points at which the investment appraisal will be refined, for example, after broad estimates of costs have been received from suppliers short-listed at the time of evaluation of Mini-Proposals. This is especially true if the suppliers have indicated that the development involves more risks, or is technically more complex, than the customer had assumed during the Feasibility Study.

5.6 Identifying Full Study and procurement activities

Before the Feasibility Options are finalized, consideration should be given to the combination of SSADM and procurement activities implied for the candidate Full Studies. In a very limited set of circumstances additional requirements elicitation in the Feasibility Study may remove the need for the subsequent Full Study to carry out SSADM Stage 1 Investigation of Current Environment and Stage 2 Business System Options. In addition, in order to draw up Outline Development Plans for the candidate projects which involve procurement, assumptions need to be made about what the procurement strategy should be, and which of the TAP procurement routes will be followed.

The decision table, Table 5.1, indicates the most common 'conditions' which may affect both the level of detail which should be put into the Feasibility Options, and the appropriateness of SSADM Stage 1 Investigation of Current Environment and Stage 2 Business System Options to subsequent Full Study projects.

The 'next activities' in the decision tables are the recommended activities (for each set of 'conditions') for the completion of the Feasibility Options and for the subsequent Full Study project, where the procurement route to be followed is assumed to be either the High Complexity or the Medium Complexity route. The position of the activities to prepare and issue the SOR and OR are shown in the sequence of 'next activities'.

	Conditions			
1	Is further requirements elicitation needed?	Y	Y	N
2	Is analysis of a current system appropriate?	Y	N	-
	Next Activities			
1	Complete 'standard' Feasibility Options	Y	N	Y
2	Complete 'extended' Feasibility Options	N	Y	N
3	SSADM Stage 1 Investigation of Current Environment	Y	N	N
4	SSADM Stage 2 Business System Options	Y	N	N
5	SSADM Stage 3 Definition of Requirements (part)	Y	Y	N
6	Prepare and issue SOR	Y	Y	N
7	SSADM Stage 3 Definition of Requirements (full)	Y	Y	N
8	Prepare and issue OR	Y	Y	Y

Table 5.1: Activities for candidate projects following TAP Medium or High Complexity route

Condition 1 in Table 5.1 tests whether further requirements elicitation is needed. If the candidate procurement project is not to be tasked with the design of the system, then the level of requirement gathered during the Feasibility Study should be sufficient for the preparation of either an SOR or an OR. For example, if the scope of the project is to be limited to replacing the hardware on which the current IT system runs, then the level of detail of the requirement elicited during the Feasibility Study should be sufficient to advertise, and issue, the first of the procurement products to suppliers, ie to commence the call for tender process. There may be a need to gather more detail about the current system platform; but there is unlikely to be any need to further analyse the requirement. In practice, the usefulness of the SOR in such procurements must be assessed on a case by case basis. The table assumes that no SOR is issued even though the procurement is following the High or Medium Complexity route.

Condition 2 in the table tests whether analysis of a current system is appropriate. It may not be appropriate for one of two reasons: either because the organization is seeking to establish a new information system, or because there are compelling business constraints which mean that the overall business benefit can only be met by not investigating the current system.

If the customer's requirement is for a new information system, then there is no current system to investigate, and SSADM Stage 1 Investigation of Current Environment is impossible. For example, where new legislation is introduced, a Government department may find that it needs a new information system to carry out its new functions.

An example of how compelling business constraints might force a candidate Full Study project to not investigate the current system is one where severe budget constraints demand that the development of the information system be restricted to the addition of a few, relatively straightforward, enhancements to the current system. In this case resources expended on a detailed investigation of the current system (SSADM Stage 1) may be unwarranted, especially if the existing system is already documented using SSADM. Another example is where the operational system is required for a limited period only, ie, two years. Here the cost to the customer of investigating the current system may be greater than the savings to be derived from the system in its two year life.

If SSADM Stage 1 is to be omitted, then, in terms of SSADM stages, this means that Stage 0 Feasibility, which concludes with selection from among Feasibility Options, is followed at the start of the Full Study by Stage 2 Business System Options, which is also concerned with option selection! In these circumstances combined Feasibility Option/Business System Option decisions should be taken at the time of selecting from among the Feasibility Options. But in order to do this safely, the Feasibility Options must present the customer with as much detail as Business System Options. In particular:

- the business models of the Feasibility Study must be taken to a lower level of detail, in which the IT system is more clearly modelled

- all the service level requirements must be specified

- sufficient detail about the risks, impact etc must be provided in support of the options to enable plans to be outlined and system boundaries to be established

- more precise investment appraisals must be provided than is normally the case with Feasibility Options.

The preparation of Feasibility Options which provide this information appears in the table as Activity 2 – Complete 'extended' Feasibility Options.

For the sake of simplicity the table does not explicitly show the next activities for projects following the Low Complexity route, though they may be implied from the table. For projects following the Low Complexity route the issue of an SOR is inappropriate. In addition, the set of 'next activities' will be very much reduced if the Feasibility Option refers to an information system development for which a package-constrained solution would be acceptable.

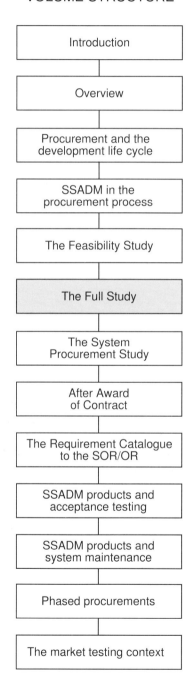

6 The Full Study

6.1 Role of the Full Study

The purpose of the Full Study is to set the scope of the proposed system and to investigate the requirements of the system in full. Alternative ways of acquiring the system are examined, taking the procurement scope defined in the Feasibility Study into account.

'Full Study' is a procurement term which refers to a part of a procurement project. The Full Study is the part of the project concerned with the detailed investigation of a proposed information system and the specification of the requirements of the system. The SSADM stages to which the Full Study equates are identified in the following section.

The system to be investigated will, in the great majority of cases, be one already examined in overview by a preceding Feasibility Study. Alternatively, the system may be a candidate application proposed in the IS Strategy, or it may be identified from within a part of the business.

If the project is to include procurement then a procurement strategy is adopted, and a decision is taken about the procurement route that will be followed. If TAP procedures are to be used and either the TAP Medium or High Complexity route is being followed, an additional objective of the Full Study is to initiate the procurement process by advertising in the Official Journal of the EC, issuing an initial requirements document called a Statement of Requirement (SOR), and evaluating and shortlisting the Mini-Proposals submitted in response to the SOR.

During the Full Study, whether it is concerned with procurement or not, the Business Case justifying the proposed expenditure is prepared and approval is sought.

6.2 SSADM stages for a procurement Full Study

This section identifies the SSADM stages which equate to the Full Study and considers the impact of procurement on the use of SSADM. Not all of the stages are appropriate in all circumstances.

Stage 1 Investigation of current environment	There are some circumstances in which the customer might choose to omit Stage 1. These are discussed in Section 5.6. However, for the great majority of procurement Full Studies all of the activities and products of SSADM Stage 1 are appropriate.
	In addition to the SSADM activities, informal contacts with suppliers initiated during the Feasibility Study may be further pursued. If the Full Study is not the result of a Feasibility Study, then informal contact with suppliers might be initiated at this stage in order to explore the procurement possibilities. Further guidance is given under 'Information gathering' in Section 5.4.
	Entries in the SSADM Requirements Catalogue should be formatted in a way which will enable them to be 'pasted' into the SOR/OR. Guidance is given in Chapter 9.
Stage 2 Business System Options (BSO)	For the great majority of procurement Full Studies all of the activities and products of SSADM Stage 2 are appropriate. However, there are some circumstances in which it may be appropriate for the customer to 'bring forward' Business System Option selection into the Feasibility Study. In short, this involves considering BSO issues at the time of selecting from among the Feasibility Options. Note that the decision to bring forward consideration of BSO issues to the Feasibility Study is always connected with the value to the customer of investigating the current system. The circumstances in which BSO issues might be considered during the Feasibility Study are described in Section 5.6. If the customer has made BSO selection during the Feasibility Study, Stage 2 is omitted during the Full Study.
	The BSOs are compiled on the basis of the requirements identified in the Requirements Catalogue. It is important in a procurement Full Study that the entries in the Requirements Catalogue are prioritized using the categories necessary for the procurement. These categories are:

- mandatory

- desirable. |

Guidance on the meaning and implications of these categories is given in Chapter 9.

Each entry in the Requirements Catalogue must be allocated one of these priority levels prior to the preparation of BSOs. BSO selection by the customer includes taking final decisions about the priority level of each requirement. If it is decided that a requirement is outside the scope of the required information system, the priority level is set to 'not required' and the entry in the Requirements Catalogue is annotated with an explanation of why the requirement was excluded.

In SSADM Stage 2 of a procurement Full Study decisions by the customer on BSO issues must be supplemented with additional decisions on the following:

- the way in which the system will be acquired, ie the combination of in-house development and procured solutions to the requirement

- the procurement strategy for the project, ie whether to undertake a 'big bang' or phased procurement

- the procurement route to be followed, ie (in TAP terms) whether to follow the High, Medium or Low Complexity route, or whether Standing Arrangements or the 'direct to ITT' route is available

- the extent to which hardware and software components can be called-off from a Framework Contract.

The time of BSO selection is the latest point by which these decisions should be taken. The decision should be informed by assumptions on these issues made during the Feasibility Study.

Stage 3 Definition of Requirements

Whenever the design of the required information system is to be developed, SSADM Stage 3 is appropriate.

SSADM Stage 3 addresses the specification of many, but not all, of the requirements which need to be included in the SOR and the OR. It addresses the functional, data and non-functional requirements; but it does not address the procurement requirements, such as the procurement timetable and the system acceptance requirements. Nor does it address the contractual requirements, such as the contract conditions that will apply.

On a procurement project – as on any other type of project – the use of SSADM should be tailored to the particular needs of the project. On a procurement project the tailored use of SSADM should take into account:

- the level of detail to which the requirement is to be specified

- the usefulness of the SSADM techniques and products as a means of specifying the requirement.

The level of detail to which the requirement should be specified is best considered at the overview level rather than on a product by product basis. Section 6.3 discusses the issues and gives guidance.

Sections 6.4 and 6.5 describe the usefulness of the Stage 3 products to, respectively, the SOR and the OR. The guidance in those sections is intended to help customers decide which of the Stage 3 products to produce in a given Full Study.

Where the procurement is following either the High or Medium Complexity route, the first of the formal customer/supplier procurement transactions is undertaken during SSADM Stage 3 with the advertisement of the procurement in the Official Journal of the EC, the issue of the SOR to suppliers who respond to the advert and the evaluation and short-listing of Mini-Proposals submitted in response to the SOR.

The SOR and its SSADM content are described in Section 6.4. The SOR should not be finalized until the completion of Step 310 Define Required System Processing and Step 320 Develop Required Data Model.

Stage 4 Technical System Options (TSO)

Where both the hardware and the software are being procured, Stage 4 is largely supplanted by the procurement activities. However, even in these circumstances, some TSO issues may need to be addressed:

- standards that will act as constraints on suppliers, eg International Standards, Open Standards, Interfacing Standards and Strategy-Level Standards

- internal design constraints, eg the physical location of existing systems with which the required system must interact

- numbers and generic types of terminals and printers.

In other instances, the consideration of TSO issues may become necessary as a result of the findings of Stage 3. For example, at the time of BSO selection it may have been assumed that in-house development would be necessary for a component of the system, because the data and processing of the component were assumed to be non-industry standard. If the investigation of the requirements of the component in Stage 3 reveals that the data and processing requirements are closer to the industry standard than had been assumed, TSOs should be prepared to assist the customer in taking a decision whether or not to procure the component.

If the hardware is being procured but the software is to be developed in-house, or vice-versa, TSOs should be prepared which describe the technical options for the components of the system which are not being procured.

Stage 5 Logical Design

In practice, all of the products of SSADM Stage 5 are a development of the products of Stage 3 to a lower level of specification detail. Therefore, the appropriateness of SSADM Stage 5 to the Full Study depends on the matters already considered concerning the tailored use of SSADM in Stage 3:

- the level of detail to which the requirement is specified

- the usefulness of the SSADM techniques and products as a means of specifying the requirement.

The level of detail to which the requirement should be specified using Stage 5 products is discussed in Section 6.3.

Section 6.5 describes the usefulness of the Stage 5 products to the OR (the SOR is issued during Stage 3), and is intended to help customers decide which of the Stage 5 products to produce in a given Full Study.

6.3 Requirements and background information

SSADM can contribute to the specification of the requirement. However in a procurement Full Study using SSADM, appropriate use must be made of SSADM to ensure that the SSADM products strike the right balance; specifying the requirement to a sufficient level of detail to enable suppliers to make good quality proposals, but not specifying the requirement in a prescriptive manner which restricts suppliers from potentially making positive contributions to the value of the procured system.

There is a significant contrast between the objectives of system analysis and design, and those of procurement.

The objective of system analysis and design is to specify the functional, data and non-functional requirements to a level of detail that can interface directly with the physical design and system build products. Indeed, it is frequently the case that some aspects of the user requirement are not discovered until this level of specification is attempted.

However, procurement has an underlying objective of managing the procurement process in such a manner as to:

- maintain the highest degree of flexibility prior to the award of contract, in order to secure the widest range of proposed solutions

- maintain the tightest control over the chosen supplier, in order to reduce customer risks.

- If the customer uses SSADM products to specify requirements in a prescriptive way, there is a very real danger of restricting opportunities for suppliers to add to the value of the procured system by the proposal of imaginative solutions. That is to say, if mandatory requirements are expressed in a rigorously specified manner the customer may lose the potential benefit of a supplier demonstrating that there is a better way of achieving the user requirement.

However, there is one type of procurement where this dichotomy of objectives is not present. This is where the supplier is required only to code and test software to the client's design, that is, the customer is seeking to procure only technical software preparation and is not seeking the proposal of a solution to his requirements.

In order that both of these contrasting objectives might be met in a solution procurement, it is necessary for great care to be taken in the way that functional, data and non-functional requirements are expressed in the SOR and in the OR using SSADM products. If at all possible, the customer should express requirements in business and operational terms only. The SSADM Requirements Catalogue is well suited to expressing requirements in these terms.

Other SSADM products encourage the customer to specify ever more detail about the entries in the Requirements Catalogue. It is recommended that these other SSADM products are included in the SOR and OR as background information only, to help suppliers to better understand the requirements, in the procurement sense.

What does 'requirements in the procurement sense' actually mean?

Firstly, requirements included in the SOR and in the OR must either be 'mandatory' or 'desirable'. If a proposal does not offer support for just one mandatory requirement it cannot be considered further. Proposals may offer support for desirable requirements, but they do not have to.

Secondly, requirements included in the SOR and in the OR – whether mandatory or desirable – must be expressed in measurable terms. This is because it must be possible, when the suppliers' proposals are evaluated, to assess whether the proposals meet each requirement, and, if they do, by how much the minimum acceptable level is surpassed. The customer needs this information in order to fully assess the quality of the proposals. The measurability of requirements is also required so that the procured solution can be tested at Acceptance Trials.

However, although all SSADM products lend themselves to the expression of requirements, not all of them lend themselves to the distinction between mandatory and desirable requirements as the procurement process requires, and not all of them enable requirements to be expressed in measurable terms.

Figure 6.1 summarizes the extent to which SSADM products can be used to express requirements in the procurement sense. Note the place of the Requirements Catalogue in this figure. It is the only SSADM product in the box marked 'requirements' – because it is the only product which is always appropriate as a means of expressing requirements in the procurement sense. Even so, entries in the Requirements Catalogue need to be formatted in a certain way to make them ideally suitable for inclusion in the procurement products – Chapter 9 gives guidance.

The second box of the figure – marked 'Requirements or Background Information' – contains the SSADM products which, under certain circumstances, may be an appropriate means of expressing requirements in the procurement sense. At other times these products provide background information only, by which the requirements are clarified.

The most common set of circumstances in which these products may be used to express requirements in the procurement sense is where the customer intends to require the chosen supplier to use them as the basis for the physical design of a bespoke solution.

Where the SSADM products are to express requirements in the procurement sense, they must be drafted to take the needs of the procurement process into account. For example, entries in the Data Catalogue should indicate

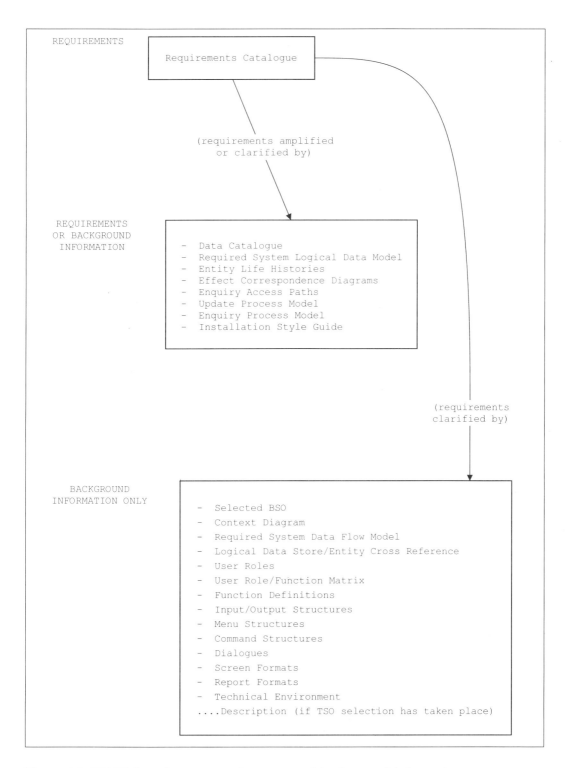

Figure 6.1: SSADM products as requirements and background information

by means of 'must' and 'should' whether the required data item is mandatory or desirable. Similarly, the Entity Descriptions and Relationship Descriptions in the Required System Logical Data Model (LDM) should use 'must' and 'should' to describe mandatory and desirable data and data relationship requirements.

Similarly, if the customer is able to specify the required functionality in terms of mandatory and desirable events and enquiries, then the ELHs, ECDs, EAPs, UPMs and EPMs may be used to express requirements in the procurement sense. However, they must be annotated in such a way that suppliers understand whether the products are expressing mandatory or desirable functional requirements.

In practical terms it may be difficult to keep the distinction between mandatory and desirable requirements clear within the SSADM products in the second box of the figure. An example is where some of the attributes of an entity are mandatory, but others desirable, or where an EAP describes navigation round both mandatory and desirable entities. The problem here is that the desirable entities may not form part of the proposed solution.

Note that these products enable requirements to be expressed only in a limited sense of the term 'measurable'. That is to say, the measurability of requirements expressed by these products is a black and white issue – either the requirement is met, or it is not. There is no sense in which one solution can exceed another in the degree to which it meets the data and processing requirements expressed in these SSADM products. For example, a required Data Item is either provided, or it is not. One supplier cannot provide the data item in a 'better' way or to a greater extent than another. Similarly, a proposed solution either includes a process which navigates round physical records equivalent to the logical entities on an EAP, or it does not.

In some cases it may be appropriate for some but not all of the SSADM products in the second box of the figure to be used to express requirements:

- the customer may wish to amplify some functional requirements by means of ECDs or EAPS, but may be happy to leave other functional requirements unamplified

- the customer may wish to amplify some data requirements by means of the LDS diagram, Entity Descriptions, Relationship Descriptions and entries in the Data Catalogue, but may be happy to leave other data requirements unamplified.

The two main risks to the customer of using the SSADM products in the second box of the figure to express requirements in the procurement sense are worth restating.

Firstly, at the time of evaluation, a proposal must be rejected if it does not meet all of the mandatory requirements. Thus, for example, a proposal must no longer be considered if it does not include a physical data item which fully implements a mandatory entry in the Data Catalogue, or if it does not include a physical data group (eg a record type or RDBMS table) which fully implements a mandatory Entity Description.

Secondly, the suppliers may have to make proposals which include tailoring or bespoke development in order to meet mandatory requirements, and they may choose to propose tailoring of a package or bespoke development in order to meet desirable requirements (because, for example, they believe that other suppliers will bid solutions which more closely fit the requirement). Such tailoring may constrain the supplier's solution technically, eg by impeding performance, and, in addition, the cost of such tailoring is likely to be passed on to the customer.

These two risks are present only in procurements under the EC Supplies Directive, or those under 'Open' or 'Restricted' Procedures under the EC Services Directive. If a procurement is under the 'Negotiated' Procedures of the EC Services Directive some flexibility in mandatory requirements can be agreed during the procurement process. Guidance on this matter is provided by CCTA in: *A Guide to Procurement within the Total Acquisition Process*.

The alternative to using the SSADM products in the second box of the figure to express requirements is to include them in the SOR/OR as background information. The SSADM products in the third box of the figure can also be included in the SOR/OR, but only to provide background information.

The advantage of including SSADM products as background information is that they can give suppliers a fuller understanding of the requirements, without constraining the way in which the suppliers can propose to meet those requirements.

6.4 The Statement of Requirement (SOR)

The purpose of the SOR is to provide for the benefit of suppliers a concise statement of the core business and operational requirements to be met by the procurement. Time spent in drafting the sections of the SOR which describe the functional, data and non-functional (quality) requirements can be reduced if the SSADM products of the Full Study, which specify these requirements, are incorporated into the SOR. There are some sections of the SOR to which no SSADM product relates. This section describes which SSADM products map onto which sections of the SOR.

The SOR is used in procurements following the High or Medium Complexity routes to initiate the formal procurement process during the Full Study so that:

- project timescales benefit from starting the procurement process during the Full Study rather than waiting until the Full Study has been completed

- suppliers benefit from low entry bid costs, because the response to the SOR is a Mini-Proposal rather than a Full Proposal. This encourages competition and helps suppliers new to the market to enter the bidding

- suppliers who are not short-listed at the Mini-Proposal stage incur only modest bid costs

- customer resources can be saved by concentrating the detailed evaluation on a small number of suppliers short-listed as a result of their Mini-Proposals (the OR is only sent to suppliers short-listed following Mini-Proposal evaluation).

The SOR is produced early in SSADM Stage 3 Definition of Requirements, after the Requirements Catalogue, Required System Data Flow Model and Required System Logical Data Model have been adjusted to take into account the selected Business System Option, and expanded to reflect all the requirements.

At this point the requirements are still expressed primarily in business terms; a lower level of detail to the specification is developed during the remainder of Stage 3. However, the SOR must not be a high-level statement of a loosely defined business problem used to initiate the procurement process, and subsequently discarded in favour of a more precise and technically more detailed requirement. In particular, all of the customer's mandatory requirements must be in the SOR. They may be amplified in the subsequent OR, but additional mandatory requirements should not be introduced between the SOR and the OR, unless the procurement is under 'Negotiated' Procedures of the EC Services Directive – see Section 6.3 for guidance in this respect.

Occasionally, highly desirable requirements may be included in the SOR, where they provide essential information to the suppliers to indicate the line along which the customer's understanding of the requirement is developing.

Table 6.1 shows the recommended structure of an SOR, and indicates which SSADM products map onto the SOR's sections and sub-sections. The remainder of this section discusses the use of the SSADM products in the SOR. For some of the products recommendations are made about how to enhance them to better suit the procurement process.

SOR section/sub-section			Relevant SSADM products
1	INTRODUCTION		None
2	BACKGROUND TO THE REQUIREMENT		
	2.1	Current Operations	Context Diagram (+) Current Physical DFM (+) Current Environment LDM (+)
	2.2	Future Developments	None
	2.3	Role of this Procurement	None
	2.4	Purpose of the System	Context Diagrams (+)
3	INFORMATION SYSTEM REQUIREMENTS		
	3.1	Functional Requirements	Requirements Catalogue Required System LDM (*) Data Catalogue (*)
	3.2	System Characteristics	Requirements Catalogue
	3.3	System Operations	Requirements Catalogue
	3.4	Design Constraints	Requirements Catalogue Overview Technical Environment Description (*)
	3.5	Service Requirements	Requirements Catalogue
4	PROCUREMENT AND CONTRACTUAL REQUIREMENTS		None
5	RESPONSE TO THE SOR		None
	BACKGROUND INFORMATION IN ANNEXES		Selected BSO (+) Required System DFM (+) User Roles (+) Logical Data Store/Entity Cross-reference (+)
*	Product may either express requirements or be included as background information only		
+	Product should not be used to express requirements		

Table 6.1: Mapping SSADM products and the SOR

(Most of the suggested enhancements are aimed at improving the traceability of requirements from the Requirements Catalogue through to the implemented system. Chapter 10 describes the requirements traceability paths which SSADM products can provide.)

In addition, in order to help managers reach their own conclusions about the usefulness of the SSADM products, an indication is given of the added value which SSADM products can bring to the procurement products. This includes describing circumstances under which the SSADM products might not be relevant, or under which the content might be reduced.

The figure shows the mapping of SSADM products onto sections of the SOR. However, this mapping does not necessarily indicate the physical location of the SSADM product in the SOR. Some SSADM products may be physically located in the section of the SOR onto which they map. This should always be so in the case of the Requirements Catalogue; and it may be so in the case of the Context Diagrams and some parts of the Overview Technical Environment Description. For the other SSADM products it is usually the case that they are most easily included in the SOR as annexes.

Context Diagram

The SOR may include a Context Diagram for the current system, and it should include a required system Context Diagram. The only circumstances where a required system Context Diagram might not be necessary is for a system with a simple behavioural model, for which the top-level Data Flow Diagram (DFD) containing a few data stores and processes can be acceptable as a replacement for a Context Diagram.

The Context Diagrams should show:

- the interfaces with other systems

- the major sources of data input to the system

- the major recipients of data from the system.

The main purpose of a Context Diagram is to summarize a system. Therefore, it must fit on one page.

A comparison of the current and required system Context Diagrams can graphically depict important changes which the required system is to introduce, eg additional interfaces, current interfaces no longer required, systems currently 'external' but which are to be integrated within the required system and new MIS outputs.

Data Flow Models (DFMs)	As Table 6.1 illustrates, the SOR may include both a Current Physical DFM and a Required System DFM. 'Current Physical DFM' is used here to refer to a set of SSADM products:

- Current Physical Data Flow Diagrams

- Input/Output Descriptions

- Elementary Process Descriptions.

The Logical DFM (the logical view of the current system from SSADM Stage 1) should not be included in the SOR.

The Current Physical DFM is primarily of benefit to the user and to the analysts, as a means of exploring the requirement. However, it can be of benefit to suppliers. It can highlight components of the current system which the required system development must not modify, eg inputs to the system from other systems. It can also be helpful in placing in context components of the current system which suppliers are expected to integrate with their solution, eg processes supported by an application package in the current system which is to be retained. Note, however, that such constraints on the required system's design must also be expressed in the Requirements Catalogue.

For the purposes of the SOR the Required System DFM should only describe the required update processing which was given a mandatory priority at BSO selection. (Mandatory enquiry processing requirements should be described by means of Requirements Catalogue entries.) The Required System DFM should provide sufficient detail to enable suppliers to scope the required system, to derive a solution and, if they are short-listed, to provide cost estimates.

The DFMs should include a set of DFDs showing the system boundary, the flows of data across it and the processes within it.

Elementary Process Descriptions (EPDs) should be prepared for each lowest level process on the DFDs. The EPDs should describe how the process is triggered, what the processing volumetrics are (for the current system), the physical nature of the process and, most importantly, they should briefly describe the business functions which the processes support.

The EPDs should serve to summarize the system processes, not describe them in detail. The way for the customer to avoid wasting resources on unnecessarily detailed EPDs is to assume that suppliers will base their response to the SOR on a Function Point estimate of the system size.

Function points are an expression of the size and complexity of a system. A function point count of a system is derived partly from the number of business entities on the LDM, and partly from the business events and the data items input to and output by the events. The EPDs should be drafted to help suppliers identify the events. Guidance on function point counting is given in the ISE Library Volume: *Estimating using Mk II Function Point Analysis*.

User tasks of less significance – in terms of updates to the system data – should be referred to in the EPDs, but should not be depicted as separate processes on the DFD. In this way, for all but the largest, most complex systems, it should be possible to restrict the set of DFDs to no more than a second level of decomposition.

I/O Descriptions should be prepared for data flows which cross the system boundary. The I/O Descriptions should describe the business data items input to and output from the processes. When describing the current system, I/O Descriptions may also describe the physical form which the I/Os currently take, if it is mandatory that they continue in that physical form.

As part of the Required System DFM, an External Entity Description should be prepared for each source of data input to the system across the system boundary, and for each recipient of data outside the boundary. Each External Entity Description should name the entity and describe its characteristics. If the entity imposes a constraint on the required system's interfaces, the nature of that constraint should be described, eg the physical location should be indicated for an existing external entity to which the required system must continue to interface.

Where the External Entity Descriptions refer to users of the system, they should do so in terms of the User Roles, ie the name of the External Entity should be the same as the name used in the User Roles.

Logical Data Models (LDMs)

As Figure 6.1 illustrates, the SOR should include the Current Environment LDM and the Required System LDM. The former provides background information to suppliers; the latter may provide background information only, or it may be used by the customer to amplify requirements in the procurement sense.

The LDMs only need to be complete enough and accurate enough to enable suppliers to scope the system. (The customer's understanding of the data requirements will be developed after issue of the SOR by means of the SSADM technique, relational data analysis.) Therefore, the Logical Data Structure (LDS) diagram of the required system need only show the business data in terms of mandatory entities (ie data groups) and the mandatory relationships between them. Desirable data requirements may be modelled for information purposes only.

For each mandatory entity an Entity Description should be provided describing:

- what the entity is
- why it is needed

- what business data items constitute the entity

- the data volumetric estimates relevant to the entity.

For each mandatory relationship on the LDS a Relationship Description may be provided. However, in order to simplify the SSADM requirements traceability paths, cardinality requirements should be described in the Entity Descriptions – Chapter 10 provides guidance on the requirements traceability paths supported by SSADM.

Desirable relationships may be referred to in the SOR, but only as 'additional information' for the benefit of suppliers.

Requirements Catalogue

Chapter 9 gives guidance on how to structure the Requirements Catalogue and on how to draft entries to better suit the procurement process. The Requirements Catalogue should cover the following topics: (for each topic, reference is given in brackets to the section of the SOR into which the entries from the Requirements Catalogue can be 'pasted')

- functional requirements (3.1)

- data requirements (3.1)

- usability requirements (3.2)

- audit (3.2)

- security (3.2)

- privacy (3.2)

- service level requirements including availability requirements (3.2)

- system control, back-up and disaster recovery requirements (3.3)

- data conversion/take-on requirements (3.4)

- EC or national standards (3.4)

- user and operations staff training requirements (3.5)
- user support requirements (3.5)
- system documentation requirements including user manual requirements (3.5)
- maintenance requirements (3.5).

Note that an entry in the Requirements Catalogue which describes a functional requirement should also describe its non-functional aspects, eg response times. In this way non-functional requirements which apply to the required functions appear in Section 3.1 of the SOR.

Non-functional requirements which apply at the system level (as opposed to applying at the level of one or more required functions) appear in Sections 3.2 and 3.3 of the SOR.

Service level requirements relating to the operation and management of the system are only relevant if the procurement is for the outsourcing of IT service provision.

Entries in the Requirements Catalogue which are amplified by means of other SSADM products should cross-refer to all of the SSADM products which amplify them. Such cross-references should be included in the SOR for the benefit of suppliers.

Only the requirements given a 'mandatory' priority at the time of BSO selection have to be included in the SOR. After the SOR has been issued, the customer may amplify the mandatory requirements – either by means of amplifying the relevant entries in the Requirements Catalogue or by means of other SSADM products – but additional mandatory requirements should not be introduced.

Desirable requirements may be included to provide suppliers with background information about the way in which the customer's understanding of the requirement is evolving.

Requirements which have been rejected should not be included in the SOR.

System sizing requirements are very important to suppliers: data volumetrics may be included in the LDM and processing volumetrics may be included in the DFM. If they are not, then it is important that they are stated in the Requirements Catalogue (in which case they appear in Sections 3.2 and 3.3 of the SOR).

Data Catalogue

The Data Catalogue provides the suppliers with further important details about the business data items. This can be particularly helpful to suppliers who may wish to propose an application package as part of their solution, because, even though the package may support the right data item types, if the format or length of the package's data items do not meet the mandatory data item requirements of the customer, then tailoring of the package will be necessary.

Overview Technical Environment Description (TED)

The TED is referred to here as an 'overview' TED, because in a procurement project much of the technical environment detail normally recorded in the TED cannot be known until after the implementation contract has been awarded.

In some Full Studies the overview TED can include requirements in the procurement sense. In others, the contents of the overview TED are suitable for inclusion in the SOR as background information only. In other procurement Full Studies there may be no need for an overview TED at all, eg where the supplier is required to simply code and test to the customer's design.

The TED is likely to express requirements if the suppliers are required to do any of the following:

- integrate their solutions with existing hardware, software and communications components of the system

- take over the running of existing hardware, software and networks

- convert hardware, software or communications components

- supply hardware.

The overview TED should indicate the physical nature and geographical location/accommodation of the hardware, software and network to a level of detail sufficient to enable the suppliers to scope the system or its conversion.

The overview TED may, by describing the need for continuity with existing systems, establish the ground for derogating from the requirement under EC/GATT rules to use standards-based specifications, where relevant and appropriate standards and conformance testing facilities are available.

For the supply of hardware, the hardware requirements in the overview TED must be in terms of numbers and generic types only, eg 30 DOS-based PCs.

Note that the requirements which these things represent should be referred to in the Requirements Catalogue and should also be referred to in the SOR as design constraints section of the SOR.

Selected Business System Option

A description of the selected BSO should be included in the SOR to provide the suppliers with background information about:

- what the selected BSO consists of (it is often an amalgam of other options put to the customer) with the major business benefits, which it is anticipated the selected option will deliver, identified

- the reasons why the option was selected

- significant reasons for not selecting other options.

User Roles

User Roles provide a succinct scoping statement for suppliers about the extent of the required on-line activity. User Roles should always be included in the SOR as background information. Brief job descriptions should be given for all the users of the system in the form of one line statements about each activity on the system which the user roles are required to undertake.

The key to success here is to ensure that the one line statements about the users' activities help the supplier to understand the business events to the greatest extent possible. However, for the purposes of the SOR, no attempt should be made to identify Functions.

Care should be taken to ensure that the User Role names are identical to the names used in Requirements Catalogue entries which describe security and access requirements.

User Catalogue	Note that this product of SSADM Stage 1 should not be included in the SOR. Its required system equivalent, the User Roles, is of far greater use to the suppliers.
Logical Data Store/Entity Cross-Reference	The Logical Data Store/Entity Cross-Reference can be provided as background information in support of the Required System DFM and the Required System LDM. Its purpose is to enable suppliers to understand what data items are read from the database and written to it by the processes described in the DFM.

6.5 The Operational Requirement (OR)

The purpose of the OR is to provide suppliers with a comprehensive statement of the business and operational requirements to be met by the proposed solutions.

In the case of procurements following the TAP High or Medium Complexity routes, the OR is issued only to suppliers who have been short-listed at the time of evaluation of the Mini-Proposals. In procurements following the Low Complexity route, the issue of the OR is the suppliers' first opportunity to consider the customer's requirement in full, though they may be familiar with the requirement to some degree if there has been informal contact with the customer.

As with the SOR, time spent in drafting the sections of the OR which describe the functional, data and non-functional (quality) requirements can be reduced, if the SSADM products of the Full Study which specify these requirements are incorporated into the OR.

There are some sections of the OR to which no SSADM product relates. This section describes which SSADM products map onto which sections of the OR.

In response to the OR suppliers submit Full Proposals. In procurements following the TAP Low Complexity route the Full Proposals are the subject of the first and only short-listing. In procurements following the High or Medium Complexity routes the Full Proposals are the subject of the second short-listing – the first short-listing is from among the Mini-Proposals received in response to the SOR.

The relationship between the OR and its predecessor, the SOR, is important in TAP. The OR should build on the SOR. Therefore, the OR should contain all the mandatory requirements stated in the SOR. Additional mandatory requirements should not be added, although existing mandatories can be amplified/elaborated. Desirable requirements should be added during the completion of the Full Study, and are included in the OR.

However, the above restrictions do not necessarily apply in every case to procurements under the 'Negotiated' procedures of the EC Services Directive.

The OR is produced after the Full Study has been completed, and after the Business Case has been made and financial approval for the procurement given.

Table 6.2 shows the recommended structure of an OR issued for a procurement which is following the TAP High or Medium Complexity route. The recommended structure of an OR for a procurement following the Low Complexity route is a sub-set of this structure. The table indicates which SSADM products map onto the OR's sections and sub-sections.

The remainder of this section provides guidance on the use of the SSADM products in the OR. Where appropriate, reference is made to guidance already provided in Section 6.4 on the products' inclusion in the SOR. The SSADM products which describe the current system should not change between issue of the SOR and issue of the OR. Therefore, for procurements following the TAP High or Medium Complexity route, where the OR is only issued to suppliers already in receipt of the SOR, the SSADM products which describe the current system may either be omitted from the OR, or copied from the SOR into the OR. The same applies to the description of the selected BSO.

Context Diagrams	The use of Context Diagrams in the OR is the same as in the SOR – see Section 6.4.
Data Flow Models (DFMs)	The use of a 'Current Physical DFM' in the OR is the same as in the SOR – see Section 6.4.

The Required System DFM should be included in the OR, but should reflect all of the requirements, both mandatory and desirable.

By the end of the Full Study, it is the Function Definitions which package the processing requirements, and, therefore, the Required System DFM becomes redundant to a great extent. Where Function Definitions are included in the OR, the main purpose of the DFDs and their supporting products is to graphically represent the system boundary, its major processes and the major cross-system boundary data flows in a way which narrative Function Definitions cannot. Time should not be expended on developing and maintaining the DFM to a third level of decomposition of the DFDs. Guidance on how to achieve this is given in Section 6.4.

The DFM should only be included in the OR if it has been maintained and if its view of the processing does not contradict the Function's view.

Logical Data Models (LDMs)	The use of the Current Environment LDM in the OR is the same as in the SOR – see Section 6.4.

The Required System LDM should be included in the OR as described in Section 6.4 for the SOR, but should reflect all of the data requirements, both mandatory and desirable. In addition, by the end of the Full Study, the LDM should take into account the results of relational data analysis. Entity sub-types should be used to express the business data, if necessary.

OR section/sub-section			Relevant SSADM products
1	INTRODUCTION		None
2	BACKGROUND TO THE REQUIREMENT		
	2.1	Current Operations	Context Diagram (+) Current Physical DFM (+) Current Environment LDM (+)
	2.2	Future Developments	None
	2.3	Role of this Procurement	None
	2.4	Purpose of the System	Context Diagrams (+)
3	INFORMATION SYSTEM REQUIREMENTS		
	3.1	Functional Requirements	Requirements Catalogue Required System LDM (*) Data Catalogue (*) Entity Life Histories (*) Effect Correspondence Diagrams (*) Enquiry Access Paths (*) Update Process Models (*) Enquiry Process Models (*)
	3.2	System Characteristics	Requirements Catalogue
	3.3	System Operations	Requirements Catalogue
	3.4	Design Constraints	Requirements Catalogue Overview Technical Environment Description (*) Installation Style Guide (*)
	3.5	Service Requirements	Requirements Catalogue

* Product may either express requirements or be included as background information only

\+ Product should not be used to express requirements

Table 6.2: Mapping SSADM products and the OR

	OR section/sub-section	Relevant SSADM products
4	PROCUREMENT AND CONTRACTUAL REQUIREMENTS	None
5	RESPONSE TO THE OR	None
	BACKGROUND INFORMATION	Selected BSO (+) Required System DFM (+) Logical Data Store/Entity Cross-reference (+) User Roles (+) Function Definitions (+) I/O Structures (+) User Role/Function Matrix (+) Dialogues (+) Specification Prototype Outputs (+) Command Structures (+) Menu Structures (+) Report Formats (+) Screen Formats (+) Selected TSO (+)
*	Product may either express requirements or be included as background information only	
+	Product should not be used to express requirements	

Table 6.2 (continued): Mapping SSADM products and the OR

If the Required System LDM is used to express requirements in the procurement sense, it must be clear from the Entity Descriptions which data items reflect mandatory requirements, and which reflect desirable requirements. Section 6.3 discusses this in detail.

Data relationships should be described in the relevant Entity Descriptions, in order to simplify the requirements traceability paths which are made available by SSADM products. Chapter 10 describes the necessary enhancements.

Requirements Catalogue — Chapter 9 gives guidance on how to structure the Requirements Catalogue and on how to draft entries to

Requirements Catalogue	Chapter 9 gives guidance on how to structure the Requirements Catalogue and on how to draft entries to better suit the procurement process. Entries from the Requirements Catalogue should be incorporated into the OR in the same way as into the SOR – see Section 6.4 above. However, both mandatory and desirable entries should be included in the OR.
	System sizing requirements are very important to suppliers. Data volumetrics may be included in the LDM and processing volumetrics may be included in the DFM or in the Function Definitions. If they are not, then it is important that they are stated in the Requirements Catalogue, in which case they appear in Sections 3.2 and 3.3 of the OR.
Data Catalogue	The Data Catalogue should be incorporated into the OR in the same way as into the SOR – see Section 6.4. However, if the Data Catalogue is used to express requirements in the procurement sense, it is important for the customer to specify which data item formats and lengths are mandatory and which are desirable, as discussed in Section 6.3.
Entity Life Histories (ELHs)	The entity life history analysis technique is of great benefit to the customer as an aid to understanding the requirement. For all but the most straightforward record-keeping systems (in which all entities have simple life histories) the customer runs the risk of not identifying all the business events in the information system if entity life history analysis is not undertaken.
	In addition, since they describe the behavioural model of the information system, ELHs are useful to suppliers as a means of scoping the requirement.
	Therefore, ELHs should be included in the OR unless the customer can justify the risk of not including them.
	However, as discussed in Section 6.3, ELHs may (unnecessarily) constrain the suppliers' solutions, if they are included in the OR as expressions of the requirement. If not included as requirements, the ELHs should be included in the OR as background information to clarify the requirements.

Effect Correspondence Diagrams (ECDs) Enquiry Access Paths (EAPs)	ECDs and EAPs show the required logical database navigation and updates for each event and enquiry. It is by developing the ECDs and EAPs that the customer uncovers the business rules which underpin the functional requirements of the system. The development of ECDs and EAPs can clarify the event and enquiry requirements, eg by identifying common enquiries. They can also help to clarify input requirements of events and enquiries. Therefore, in order to avoid the risk of not fully understanding the requirement, the customer should develop ECDs and EAPs for all systems, except those with very simple behavioural models, ie except the most functionally straightforward information systems. However, as discussed in Section 6.3, ECDs and EAPs may (unnecessarily) constrain the suppliers' solutions in the same way as ELHs. If this is likely, the ECDs and EAPs should be included in the OR as background information only to clarify the requirements.
Update Process Models (UPMs) Enquiry Process Models (EPMs)	UPMs and EPMs, collectively called 'process models', are logical descriptions of the processing requirements, for communication with the logical database, of the events and enquiries. UPMs are derived from ELHs and ECDs, and EPMs are derived from EAPs and I/O Structures. Therefore, UPMs and EPMs can only be included in an OR in which ELHs, ECDs, EAPs and I/O Structures are also included. The argument in favour of the customer developing the process models, even if only first cut versions which are not included in the OR, is that their development helps to elicit aspects of the requirement which may otherwise be overlooked. For example, although ECDs and EAPs contain selections and iterations, the conditions which control them are not specified until development of the UPMs and EPMs. The process models complete the definition of the conceptual model of the required system. Without them the customer cannot be sure that the requirement has been fully understood.

Experience has shown that UPMs and EPMs can be of as much benefit to developers using a 4GL as to developers using a 3GL. Therefore, whether produced by the customer as part of the Requirements Specification, or produced by the supplier under the implementation contract, the process models remain valuable (to both customer and suppliers) throughout the operational life of the system, irrespective of the maintenance environment.

However, as discussed in Section 6.3, UPMs and EPMs may (unnecessarily) constrain the suppliers' solutions in the same way as ELHs. If this is likely, the UPMs and EPMs should be included in the OR as background information only to clarify the requirements.

In certain circumstances the customer may choose not to develop the process models, eg where the customer anticipates that suppliers will be able to propose acceptable solutions without the processing requirements specified to the level of detail of process models. This is most likely to be the case where suppliers are able to offer application packages which can meet the business requirements with only a moderate amount of tailoring.

Another reason for the customer not to develop process models is if he would prefer to pay for their development by the chosen supplier under the implementation contract, or, if one or more System Procurement Study is to be funded, the customer may wish to fund development of 'first cut' process models under the SPS contract. The SPS is discussed in more detail in Chapter 7.

Overview Technical Environment Description (TED)	An overview TED should be included in the same way as in the SOR – see Section 6.4.
Installation Style Guide	The customer's Installation Style Guide sets the standard for the user environment. This includes ergonomic details, such as the siting of equipment and system based requirements, such as the style that is to be used for dialogues and reports.

procurement sense, or as background information for suppliers.

When the Installation Style Guide is used to express requirements, it should only amplify entries in the Requirements Catalogue which state the mandatory and desirable Human Computer Interface (HCI) requirements. The Installation Style Guide should be annotated appropriately to distinguish the parts which amplify mandatory requirements from the parts which amplify desirable requirements.

The Installation Style Guide can be particularly useful to customers, who are procuring more than one information system, as a means of ensuring that the solutions do not result in an overly varied set of interface styles.

However, the risk to the customer of including the Installation Style Guide (or parts of it) in the OR as a means of expressing requirements in the procurement sense is that it may constrain suppliers' solutions to the customer's overall disadvantage. For example, a supplier may not feel able to propose a solution which would more than meet the business requirement, simply because the application package at the heart of the solution could not meet one or two mandatory HCI requirements without significant (and therefore costly) tailoring.

In some procurements, therefore, the Installation Style Guide is included in the OR, but only as background information; in other procurements the customer excludes the Installation Style Guide from the OR (in which case the suppliers need to take into account only those entries in the Requirements Catalogue which describe the HCI requirements).

Selected Business System Option	The selected BSO should be included in the OR in the same way as in the SOR – see Section 6.4.

Logical Data Store/Entity Cross-Reference	The Logical Data Store/Entity Cross-Reference should be included in the OR in the same way as in the SOR – see Section 6.4.
User Roles	User Roles should be included in the OR in the same way as in the SOR – see Section 6.4. If the OR describes user operating manual requirements, these should be expressed in terms of User Roles.
Function Definitions	Function Definitions provide the users' view of the system processing. In addition, for each function, the Function Definitions package together:

- the events and enquiries

- the inputs and outputs.

Each Function Definition includes a narrative description of the function, and is supported by I/O Structures and by the SSADM products which specify the event and enquiry processing.

In a procurement Full Study the development of Function Definitions and their supporting products is as beneficial to the customer's understanding of the requirement as in a non-procurement SSADM project.

However, in the procurement Full Study the customer should not use the Function Definitions as a means of expressing requirements in the procurement sense, because to do so might unnecessarily constrain the suppliers' proposals in the same way as ELHs (see above). For example, if the sequence of input and output data items in the I/O Structure is made a mandatory requirement, then a supplier with an application package which meets the business requirement (and which, maybe in addition, offers significant unexpected benefits to the customer) may nevertheless be obliged to tailor the package if the package's I/O sequence differs from the requirement. The expense of this tailoring is likely to be passed on to the customer.

Furthermore, an application package which meets the business requirement might not 'bundle together' the on-line functionality in the same way as the customer has chosen to in the Function Definitions. Therefore, even if the Function Definition is categorized as only a desirable requirement, the customer may find that the proposed solution does not contain functions which correspond one-to-one with the SSADM functions.

In procurement Full Studies, therefore, Function Definitions should be included in the OR as background information to clarify the functional requirements in the Requirements Catalogue.

In 'standard' SSADM the Function Definition is also used to record the service level requirements which are specific to the function. In a procurement Full Study the function-specific service level requirements should be documented with the appropriate functional requirement entry in the Requirements Catalogue. This avoids having measurable 'requirements' in a product which is providing background information only.

Given that all functional requirements are identified in the Requirements Catalogue, an issue for the customer to resolve is how much detail to put in the Function Definitions. The ideal is for the customer to include only that amount of detail sufficient to clarify the functional requirements needed by the suppliers to formulate acceptable proposals.

However, the amount of detail needed by the suppliers to do this depends on what the customer is seeking to procure. At one extreme, if the customer is seeking to procure the code and test of the customer's own logical design, then detailed Function Definitions (supported by detailed event and enquiry products such as ECDs and UPMs) should be included in the OR. The level of detail must be sufficient to enable a supplier to complete the Function Definitions as part of physical process specification in SSADM Stage 6. Note, however, that it may be difficult for the customer to mandate the use of SSADM by the supplier to complete the design – advice on this issue should be sought from the customer's procurement specialists.

At the other extreme, where the customer is confident that acceptable proposals will be received which are based to a large extent on application package(s), the Function Definitions may be left at a higher level of detail, or, for functional requirements which are 'industry standard', it may not be necessary to amplify the requirements by means of Function Definitions at all.

Although it is normally the case that Function Definitions should not express requirements in the procurement sense, some components of a function may relate to mandatory or desirable requirements. For example, if one of the function's events, or one of its inputs or outputs has been specified as mandatory, the reference to the event or I/O in the Function Definition should be appropriately annotated. This will help to ensure that the suppliers' proposed solutions for the functional requirement take the mandatory event or I/O into account.

In a procurement Full Study, Function Definitions may be enhanced in order to complete the requirements traceability paths in readiness for acceptance testing. The cross-references from Function Definitions to other SSADM products to provide requirements traceability paths are described in Chapter 10.

When the system is operational and the customer wishes to express maintenance requirements, Function Definitions should continue to be the medium by which the user views the system (and the supplier may find it helpful to maintain a map of how the user's functions map onto the internal design of the system).

The Function Definition specifies which events and enquiries make up a function, and what their frequencies are. I/O Structures indicate the order in which the events occur but it can be helpful, for example, for suppliers sizing a system, to summarize events within a function.

Consider the following (extreme case) example. An off-line function consists of 3 events and 1 enquiry. The Function Definition simply lists them thus:

Event 14 Change of address. Frequency = 200

Event 61 Deposit transfer. Frequency = 40

Event 32 Receipt of credit. Frequency = 400

Enquiry 6 Credit limit summary. Frequency = 1

What this does not tell the supplier is that all of the receipt of credits (Event 32) must be dealt with first, then all of the deposit transfers (Event 61), then all of the changes of address (Event 14), and the function must conclude with a credit limit summary (Enquiry 6).

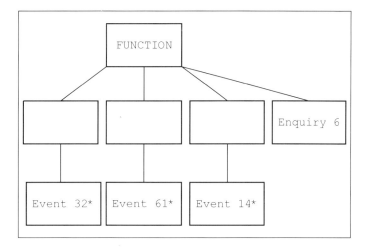

Figure 6.2

A simple 'function structure' diagram using Jackson notation should be prepared for the benefit of the supplier (see Figure 6.2). The box at the top of the diagram should represent the function; each of the boxes in the structure should represent an occurrence of an event or enquiry.

I/O Structures	I/O Structures support the Function Definitions by describing the input and output requirements of the functions.
	For the same reasons given for Function Definitions, the I/O Structures should normally be included in the OR as background information.
	However, in certain circumstances, I/O Structures may represent amplifications of a requirement in the procurement sense. For example, where an off-line function is required to read a file from another system, the data content and structure of the input is constrained by the design of the external system. In this case the Requirements Catalogue and, possibly, the overview TED should refer to the design constraint, and the I/O Structure should be used to amplify the mandatory requirement by describing the sequences and iteration in the data flow. In this case, the I/O Structure should be annotated appropriately, and cross-references should be provided linking the entry in the Requirements Catalogue to the I/O Structure.
	Because it is not normally the case that I/O Structures represent requirements in the procurement sense, I/O Structures appear under 'Background Information' in Table 6.2.
User Role/Function Matrix	Whenever Function Definitions and User Roles are included in the OR, a User Role/Function Matrix should also be provided as background information.
	The matrix is a powerful scoping document. It gives suppliers an overview of the on-line functionality of the system in terms of:

- what functions are required on-line

- who must be able to access the on-line functions

- who must be prevented from accessing the on-line functions.

Dialogues

For users of an information system there is normally no SSADM technique more helpful in eliciting requirements in SSADM Stage 3 than specification prototyping. If this technique has been used, then the users' dialogue requirements may have become fairly clear for all the functions visited during specification prototyping. At the end of SSADM Stage 3, for functions not visited during specification prototyping, the users' dialogue requirements will be expressed simply as I/O Structures.

In SSADM Stage 5 the dialogue requirements are defined more precisely by the development of Dialogues. However, the inclusion in the OR of Dialogues – or of any other form of dialogue specifications – as expressions of requirement may unnecessarily constrain the suppliers' solutions. For example, a supplier may wish to propose an application package which meets the business requirement for a dialogue but which does not meet exactly the man–machine exchanges specified in a Dialogue. Alternatively by making best use of their proposed development tools, suppliers may be able to provide dialogues which meet the business requirement – or even surpass it – but which do not meet the detailed requirements of the customer's Dialogues.

The risk of unnecessarily restricting suppliers' solutions by using Dialogues to express requirements applies in all procurements of a design, from those in which the customer is seeking to procure the code and test of the customer's own logical design, to those procurements in which the customer expects the solutions to be based on application packages.

Therefore, if SSADM Dialogues – or any other dialogue specification products – are included in the OR, they should be included as background information only to clarify the dialogue requirements.

In some instances the customer may choose not to develop SSADM Dialogues (or any other dialogue specification product) at all. For example, if a System Procurement Study is to be undertaken, the customer may prefer to have the Dialogues developed under the SPS contract. This issue is discussed in Chapter 7.

In other instances, if it proves difficult for the users to agree dialogue specifications during the Full Study, then the dialogue requirements could be included in the Requirements Catalogue but not developed by means of SSADM Dialogues.

If SSADM Dialogues are omitted from the OR, sufficient information about the dialogue requirements should be included in the OR to enable the suppliers to scope the external design of the system. This information should include:

- dialogue numbers

- dialogue relative complexity

- dialogue navigation requirements

- dialogue level 'help' requirements.

Specification Prototype Outputs

If the customer has carried out specification prototyping in SSADM Stage 3, some of the output from the technique may be included in the OR as background information to clarify the dialogue requirements.

A wide variety of prototyping software tools are available, and the outputs from them which might be appropriate for inclusion in the OR are equally varied. However, the types of product which may be included in the OR are:

- screen formats

- first cut Graphical User Interface (GUI) designs

- prototype pathways

- prototyping reports.

Command Structures

Command Structures specify the requirements for navigation around the menus and dialogues at the completion of the required dialogues. Command Structures should be included in the OR as background information to support the Function Definitions for on-line functions.

Menu Structures	Menu Structures may be included in the OR as background information.
Report Formats	Report Formats may be developed during the Full Study to amplify the entries in the Requirements Catalogue which express output requirements.
	In the majority of cases Report Formats should be included in the OR as background information. In exceptional circumstances, for example, where legislation stipulates the physical layout of a form that the system must print, it may be necessary to annotate the Report Format appropriately and cross-refer it to the mandatory entry in the Requirements Catalogue which refers to the design constraint imposed by the legislation.
Screen Formats	Screen Formats may be developed during the Full Study to support the Dialogues. If the Screen Formats are included in the OR, they should be used to provide background information only.
Selected Technical System Option (TSO)	The circumstances in which the customer may prepare TSOs during a procurement Full Study are described in Section 6.2.
	If any form of TSO selection has taken place during the Full Study, a description of the selected TSO should be included in the OR as background information. If any aspects of the selected TSO introduce technical environment requirements, the technical environment requirements should be expressed in the Requirements Catalogue and in the overview TED.
	'Selected TSO' here refers to matters such as the customer's chosen development tool, where the procurement is of hardware only, or the customer's chosen operational platform, where the procurement is of software only.
	The 'selected TSO' may also include decisions about standards to be applied.

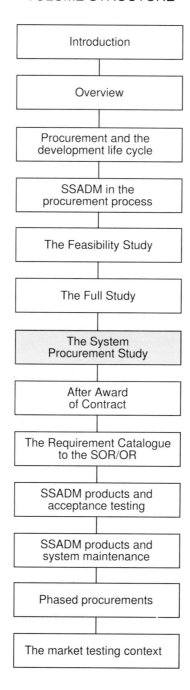

7 The System Procurement Study

The System Procurement Study (SPS) is a study funded by the customer and carried out by a supplier, or suppliers, to investigate and develop aspects of the supplier's proposed solution, and to present the results for evaluation by the customer.

SPSs are a feature only of procurements following the TAP High Complexity route, for lower risk procurements the cost to the customer of funding SPSs is unwarranted.

Where SSADM has been used during the Full Study to specify the requirements, the customer may wish to enter into an SPS contract, with up to three suppliers, under which SSADM design products are developed as part of the SPS deliverables. This chapter discusses these issues.

7.1 The role of the SPS

The purpose of the SPS is to reduce risk for both the customer and the supplier. This section highlights the principal ways in which the SPS reduces risks.

In terms of customer/supplier procurement transactions the SPS serves as a type of 'tender response' transaction, in which the supplier delivers to the customer a very detailed technical proposal and a comprehensive delivery plan. Informed by these deliverables the customer is able to commence the 'supplier selection' transactions.

The purpose of the SPS is to reduce risk. Risks to the supplier are reduced, because during the SPS the supplier has an opportunity to:

- establish a full understanding of the requirement

- develop and test the proposed solution

- prepare an implementation plan in which the supplier and the customer have full confidence.

Risks to the customer are reduced because:

- the SPS deliverables enable the customer to evaluate in detail an inherently complex system solution

- co-operation between the customer and the supplier during the SPS enables the customer to establish confidence in the supplier's ability to deliver what is being proposed.

The role of SSADM in the SPS is to act as a common language between the customer and supplier. By means of SSADM products in the OR the customer expresses the functional, data and non-functional requirements, and provides additional background information in support of the requirements. By means of the SSADM products in the SPS deliverables the supplier describes to the customer, in terms which the customer understands, aspects of the logical and physical design of the proposed solution.

7.2 SSADM stages in the SPS

The SSADM stages which may equate to an SPS are:

- Stage 5 Logical Design

- Stage 6 Physical Design.

Where the supplier agrees under the SPS contract to carry out design activities using SSADM, Stage 6 always applies. However, the applicability of Stage 5 to the SPS is not so clear cut. The customer may already have completed Stage 5 as part of the Full Study, or the customer may not wish to fund Stage 5 activities during the SPS.

Figure 7.1 summarizes the possible scenarios of SSADM stages leading up to and carried out during the SPS. All of the scenarios 'pick the story up', as it were, with Stage 3 Definition of Requirements, which is always undertaken by the customer during the Full Study. All the scenarios end with Stage 6, undertaken by the supplier during the SPS.

The reader should remember throughout this chapter that references to the SSADM activities of the supplier

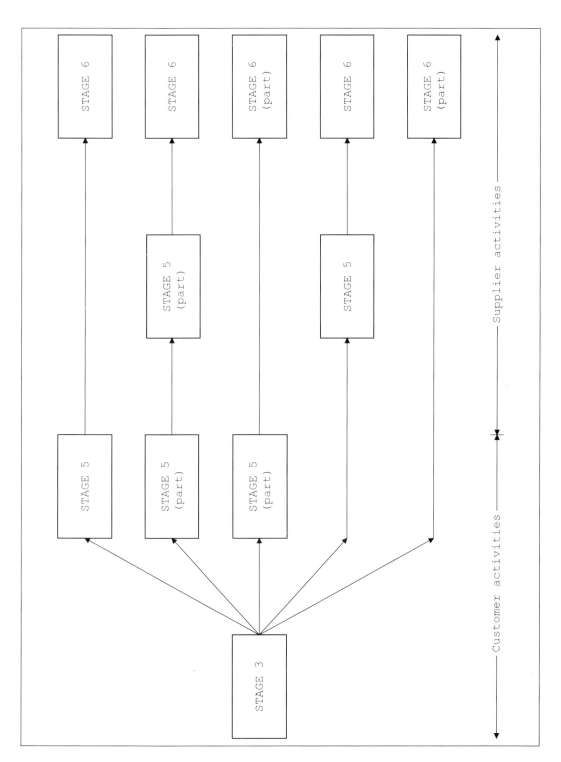

Figure 7.1: SSADM stages in the SPS

are, except where otherwise stated, the supplier's SSADM activities under the SPS contract. They are not the activities of the chosen supplier under the implementation contract.

7.2.1 Stage 5 Logical Design

As Figure 7.1 illustrates, Stage 5 may or may not have been undertaken by the customer during the Full Study. Chapter 6 describes the advantages and disadvantages to the customer of undertaking Stage 5 during the Full Study. To further complicate the possible scenarios, the customer may have chosen to prepare some, but not all, of the Stage 5 products during the Full Study.

The uppermost scenario depicts the straightforward case, where the customer has prepared all of the SSADM products of Stage 5 during the Full Study, and the supplier uses them to develop a 'first cut' SSADM physical design in the SSADM Stage 6 of his SPS.

The remainder of the figure depicts the less straightforward scenarios, in which the customer has either prepared none or only some of the Stage 5 products. In these scenarios, typical reasons for the customer choosing to have first cut Stage 5 products developed by the supplier during the SPS are:

- to help the supplier understand a complex requirement

- to ensure that the supplier provides proof that he understands the requirement in a medium other than the development environment

- to more completely involve the supplier with the user community

- to assess the supplier's proficiency

- to give the supplier an opportunity to demonstrate how the Stage 5 products will be transported into a maintenance environment, for example, into a repository or data dictionary system

- to give the supplier an opportunity to demonstrate how, under the implementation contract, the Stage 5 products will be taken forward into Stage 6 Physical Design, and thence implemented

- in a phased procurement to assess how well the supplier's Stage 5 products will integrate with the Stage 5 products of other increments.

Although the SPS is funded, it is unreasonable to expect the supplier to complete a full Stage 5. For example, the supplier may be expected during the SPS to develop first cut versions of some, but not all, of the required process models, and SSADM Dialogues for some but not all of the required on-line functions.

Furthermore, the process models and dialogues developed during the SPS, are unlikely to be the versions which will be used by the supplier as input to the physical design activities under the implementation contract.

In short, it is important for the customer to remember that the supplier will need to revisit SSADM Stage 5 if he is awarded the implementation contract.

The reasons for the customer choosing not to have Stage 5 products developed during the SPS are:

- the implementation contract with the chosen supplier is not going to include the development of an SSADM design

- the dialogue requirements have been left (deliberately or otherwise) at a high level, and the external design of the system is better suited to a rapid application development approach using the supplier's preferred development tools, or implementation by the external design of an application package

- the need for the procurement to follow the High Complexity route stems from high risk or high complexity aspects of the development other than the design, eg where the major risk concerns the robustness of the chosen hardware

- there is a need to reduce the cost of the SPS.

With regard to the final point in this list, if the customer does not take the opportunity of the SPS to assess the supplier's ability to carry out Stage 5, there is a risk that the supplier may not complete Stage 5 satisfactorily under the implementation contract.

7.2.2 Stage 6 Physical Design

As Figure 7.1 illustrates, the extent to which a supplier can develop Stage 6 products during the SPS depends in part on the availability of Stage 5 products, which may have been produced by the customer and/or by the supplier himself, or they may not have been produced at all.

If the Stage 5 products are not produced, then not all of the products of Stage 6 can be developed. The lowermost scenario in the diagram depicts this. Most importantly, without UPMs and EPMs from Stage 5, the Process Data Interface in Stage 6 cannot be developed. This means that the physical processes which access the database will each require a physical specification (ie program specification).

Also, if the Stage 5 products are not available, the products of SSADM physical process specification need to be adapted to bridge the gap between the SSADM specification products and the physical design products. For example, the FCIM must document not only which components of the physical design make up a function's dialogue, but also how those physical components relate to the function's I/O Structure, which is the specification product from which the physical dialogue is likely to have been derived, at least in part.

The format of the Stage 6 products depends on the supplier's chosen development environment. Therefore, it should be a matter for agreement between the customer and the supplier which Stage 6 products will be developed, and what format they will take. This applies to both the SPS contract and the implementation contract.

It is unreasonable for the customer to expect the supplier to complete a full Stage 6 during the SPS, that is, to develop a complete physical design of the proposed solution. The supplier will need to revisit Stage 6 if awarded the implementation contract.

What it is reasonable for the customer to expect is that sufficient products will be developed during the SPS to enable the customer to assess the supplier's ability to develop a physical design using SSADM, and, where appropriate, to demonstrate how the logical and physical designs fit together (an important consideration for the maintenance of the system in the future).

Where the proposed solution is based in part or in full on an existing application package, an essential quality criterion of the supplier's Stage 6 products is that they document how the design of the package, and the physical files and processes which implement the design, would be integrated with the bespoke design.

It is usually the case that SPS suppliers are anxious to develop prototype software as part of the deliverables from the SPS. The customer is better able to assess the abilities of the supplier, if the prototyped functionality is demonstrably derived from the first cut SSADM physical design.

7.3 SSADM products from the SPS

The main reason for the customer to fund the development of SSADM products during an SPS is to assess the supplier's ability to deliver the system that the customer requires. However, this will also enable the supplier to demonstrate the standard and format to which SSADM design products will be prepared, and the use which the supplier will make of the SSADM products if awarded the implementation contract.

Table 7.1 lists the possible SSADM products of an SPS.

STAGE 5 PRODUCTS	STAGE 6 PRODUCTS
Dialogues	Application Development Standards
Command Structures	Physical Design Strategy
Menu Structures	Physical Data Design
Update Process Models	Function Component Implementation Map
Enquiry Process Models	Screen Formats
	Report Formats
	Physical Environment Specification
SPECIFICATION PRODUCTS (with resolutions added) Requirements Catalogue Required System Data Flow Model	

Table 7.1: SSADM products from the SPS

Stage 5 Products — Reference is made in Chapter 6 to the (possible) preparation of these products by the customer during the Full Study. The following remarks apply to the preparation of first cut versions by the supplier during the SPS.

Dialogues — Where the procurement is of a fully bespoke solution or of a solution based on a package but including a significant amount of tailoring, the supplier is likely to want to place emphasis on prototyping at least parts of the bespoke functionality during the SPS. This is because the supplier will be anxious to assure himself that he can develop the functionality satisfactorily, and also because functionality which has been developed into demonstrable software is easier for the customer to evaluate.

However, even where prototyping is undertaken, there may still be a role for SSADM Dialogues to play:

- where the customer intends SSADM Dialogue design to be an activity under the implementation contract, the supplier may need to demonstrate his proficiency in the technique by developing dialogues during the SPS

- where the supplier does not have the time or resources to prototype the whole of the proposed external design, SSADM Dialogues provide an alternative way of describing the non-prototyped dialogues

- where the supplier proposes a user interface which does not lend itself to fast prototyping, for example, a user interface developed using a 3GL, the supplier may develop SSADM Dialogues as a means of feeding back to the customer how he proposes to meet the dialogue requirements.

For the purposes of evaluating the SPS deliverables, it is helpful to the customer if some or all of the SSADM Dialogues for which prototype dialogues have not been developed are submitted to the customer in combination with screens which have been developed, for example, by means of a simple screen-painting tool.

Command Structures	Command Structures define, for each User Role, what navigation around the menus and dialogues is available when a user completes or terminates a dialogue.

In the SPS the Command Structures define how the supplier proposes to bundle together the system's on-line functionality. The customer needs to evaluate how well the proposed 'bundles' of functionality map onto the user's way of working.

If the Requirements Specification has not addressed the requirements concerning navigation between functions in detail, then the SPS offers an opportunity for the customer and the supplier to come to an agreed interpretation of the relevant requirements.

Menu Structures	Menu Structures take the form of diagrams depicting, for each User Role, how it is proposed that the menus and dialogues will interrelate. For some development environments, such as those which present Graphical User Interfaces, Menu Structures are inappropriate.
Update Process Models (UPMs) Enquiry Process Models (EPMs)	Where process models are prepared as part of the SPS product set, customers should apply the quality criteria as described in the SSADM manuals as strictly as possible. Minor variations – such as abbreviations – may be permitted in the syntax of the operations, though not if they introduce any of the syntax of the proposed development language. In addition, if the proposed development environment is beyond doubt (for example, 4GL rather than 3GL), it is sensible for the supplier to tailor the operations so that those which are inappropriate are omitted.

Stage 6 Products

Application Development Standards	On procurements where the solution includes a significant amount of bespoke design and the customer requires the design to be transportable, it is important for the supplier to describe the proposed approach to physical design. In SSADM the Application Development Standards define the standards to be used throughout the application design, construction and testing. Application Development Standards should describe the following: • **Application Naming Standards**. The standards which the supplier proposes to use for naming bespoke elements of the system, for example, data item names, screen names, program names • **Application Style Guide**. The standards agreed between the customer and the supplier for the user environment, for example, ergonomic details, style issues relating to screen and report layouts, and size and positioning of items such as headings and help text. The Application Style Guide should take into account the customer's Installation Style Guide if the latter has been included in the OR (see Section 6.5 for details)

- **Physical Environment Classification.** The characteristics of the environment in which the supplier proposes to implement the solution, together with, if the environments are different, the characteristics of the proposed development environment

- **Physical Design Strategy.** A record of the planned approach to all aspects of physical implementation – this can cover a wide range of topics, about which more detail is given below.

The supplier's proposed Application Development Standards provide the customer with an overall understanding of the supplier's proposed physical design and of his approach to design and development activities. This is particularly useful where the customer requires that the procured design should be transportable, since an important part of assessing the risk associated with the supplier's proposed solution is to assess how easily the customer, or a third-party supplier on behalf of the customer, could maintain or adapt the system using the SPS supplier's design.

Physical Design Strategy (PDS)

In SSADM the PDS forms part of the Application Development Standards. The PDS is a record of the planned approach for all aspects of physical implementation and, therefore, should be developed in some detail during the SPS. By the time of evaluation of the SPS deliverables, the PDS should cover the following topics:

- identification of the major data handling, performance and processing characteristics of the proposed software

- description of the standards for all types of implementation object

- description of the techniques (SSADM and others) which will be used, and the documentation standards

- the basis upon which decisions will be taken about whether to implement processes using a procedural or non-procedural language, or, if an object-based approach is to be followed, a description of the approach that will be taken to packaging data and code into objects

- a description of the supplier's approach to the SSADM technique of physical process specification (with reference to relevant standards), including program specifications, procedural process designs, macros and batch run flow documentation

- in the case of a distributed system, the approach that will be taken to mapping the design onto the distributed elements

- the criteria which will be used to resolve the contradictory requirements of the design. For example, if it becomes apparent that a trade-off is needed between, for example, a response time which is a desirable requirement and a one-to-one mapping between the logical and physical data design

- the way in which the Function Component Implementation Map (FCIM) will be implemented. That is, what form it will take, and what software tool will be used to maintain it

- the use which will be made of a Data Dictionary System (DDS), especially where the DDS is capable of holding and linking together logical as well as physical products

- if the physical data design is unlikely to map one-to-one with the logical data design, the supplier should describe how the Process Data Interface (PDI) will be documented.

The Application Development Standards, the FCIM and the PDI are examples of SSADM products about which the method deliberately avoids being prescriptive, in order to maximize its relevance to differing technical environments. It is the supplier who makes proposals concerning these Stage 6 products, what they will contain and how (in what format) they will be produced.

It is for the customer to decide whether the supplier's proposals are acceptable, and to eventually agree Product Descriptions for the Stage 6 products with the supplier.

Physical Data Design

The Physical Data Design is an implementation-specific data design. Whenever the procurement includes procuring a bespoke design, or an element of bespoke design, it is nearly always the case that a Physical Data Design must be produced by the supplier. Only where the Physical Data Design of an application package already meets the customer's data requirements and performance objectives is no physical data design necessary on the part of the supplier.

From the point of view of the supplier, the primary purpose of the Physical Data Design in the SPS should be to serve as the design of a demonstration database which supports the prototyped software.

Equally important from the customer's point of view is that the first cut Physical Data Design included in the SPS should indicate how the supplier would approach the design of the system's underlying data, and what documentation standards would apply. This information is needed by the customer in order to assess how transportable the supplier's design would be.

SSADM does not prescribe the content or format of a Physical Data Design. However, the Physical Data Design in the SPS should describe how the LDM will be implemented in the proposed physical system. It should also indicate the extent to which it is proposed that the Physical Data Design will diverge from the logical view of the LDM, together with reasons for the divergence.

Where the requirement is to be met by a distributed system, the Physical Data Design of the SPS should also demonstrate the supplier's proposed standard for documenting the distribution of the data.

SSADM refers to the completion of what it calls Space Estimation Forms after the Physical Data Design is complete. SSADM does not prescribe the content and format of Space Estimation Forms, but they, or their equivalents, are used by database designers to estimate the total size of the physical database from the Physical Data Design. The customer may gain an insight into the competence of the SPS supplier by asking for copies of the Space Estimation Forms, or their equivalent.

Function Component Implementation Map (FCIM)

The FCIM specifies the implemented objects (eg programs, screens, sort routines) which meet the processing requirements of the components of the Function Definitions.

The Physical Design Strategy in the SPS should describe how the supplier proposes to implement the FCIM. Where it is important to the customer that the design should be transportable, for example, to a third-party supplier for maintenance of the system, the supplier should develop some elements of the FCIM during the SPS using the approach proposed in the Physical Design Strategy.

The production of at least some elements of the FCIM is not an unnecessary overhead to the supplier. If the supplier agrees to use SSADM to carry out physical design, a first cut FCIM is the obvious means by which the supplier documents, for the benefit of the customer, how specific functions have been implemented in the prototyped functionality of the SPS. For example, it can describe which functionality is provided by means of package software.

In addition, the FCIM produced during the SPS demonstrates the supplier's proposed standard for documenting the FCIM in Stage 6 under the implementation contract.

If the first cut FCIM is supported, where appropriate, by copies of what SSADM refers to as Timing Estimation Forms, the customer is able to evaluate more thoroughly the ability of the SPS supplier to deliver the proposed solution. SSADM does not prescribe the content and format of Timing Estimation Forms, but they, or their equivalents, are used by database designers to estimate the performance of processes against the operational database. By examining the supplier's Timing Estimation

	Forms the customer is able to carry out a risk assessment of the supplier's performance estimates.
Screen Formats	A Screen Format shows how items will be presented on the screen to the user. Where the supplier develops screens for demonstration purposes during the SPS, the relevant Screen Formats should be included in the SPS deliverables. The Screen Formats should conform to the user interface standards in the Application Style Guide, or its equivalent, and to the Installation Style Guide if this has been included in the OR.
	An assessment of the Screen Formats by the customer (for example, how easy are they to understand) is particularly important where the customer requires the design of the system to be transportable.
	The development of demonstration screens by the supplier during the SPS may help to clarify the dialogue requirements. However, the agreement to Screen Formats during the SPS should not be taken as an opportunity to add to the dialogue requirements.
Report Formats	In some instances Report Formats may have been included in the OR as background information. In other instances Report Formats may have been included in the OR as expressions of the requirement, for example, where predefined printer widths are a design constraint which the supplier must meet.
	If the development of demonstration software during the SPS is the first opportunity for the user to consider in detail the layout of reports, then, as with Screen Formats, it is important that the SPS activities are not taken as an opportunity to introduce additional requirements.
Physical Environment Specification	The Physical Environment Specification specifies the supplier's proposed hardware and software products and services, and the configuration in which they are to be supplied, commissioned and made available for implementation.
	SSADM does not prescribe the content or format of a Physical Environment Specification. However, one of the advantages to the supplier is that the SPS presents an

opportunity for the Physical Environment Specification to include the optimum hardware and software products/versions.

Specification Products (with resolutions added)	On procurements using SSADM a key to the successful management of the customer/supplier interface lies in the use of SSADM as a lingua franca between the customer and the supplier. Where the Requirements Specification has been developed using SSADM, the customer's perception of the future system is likely to be in terms of the (most easily understood) SSADM products, such as the Requirements Catalogue and the Required System Data Flow Model. Suppliers should take the opportunity in the SPS to describe the proposed solution using, wherever possible, the SSADM products with which the customer is most familiar.
Requirements Catalogue	Entries in the Requirements Catalogue are not complete until the 'resolution' field has been annotated with the products which 'resolve' the requirement. At the time of issue of the OR to suppliers, the resolution of entries in the Requirements Catalogue should, wherever possible, refer to SSADM products which amplify or clarify the requirement, and these cross-references should be included in the OR.
	As part of the SPS deliverables the supplier should provide the customer with a copy of the Requirements Catalogue in which, against each entry, the supplier includes in the 'resolution' field statements on the conformance of the proposed solution to the requirement. Such statements should refer, for example, to proposed physical processes which would implement the functional requirements, and to proposed files which would implement data requirements. In addition, the statements should make clear by how much the proposed solution would exceed the non-functional requirements.
Required System Data Flow Model (DFM)	As part of the SPS deliverables the supplier should provide the customer with a version of the DFM annotated to describe the proposed system structure. The supplier should describe, for at least those processes which are going to be technically difficult to implement or least likely to yield the desired benefits, the clerical

and management procedures which would be needed to support the process, and the supplier's technical approach to meeting the processing requirements.

In addition, the supplier should use the DFM to highlight any mismatch between the Business System Option's view of the extent to which IT would support the required processes, and the extent of the IT support which the supplier is proposing. For example, consider the sub-set of a DFD shown in Figure 7.2, relating to the customer's requirement for a data capture facility.

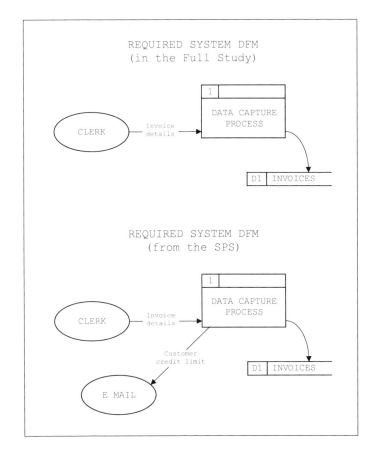

Figure 7.2

The part of the figure entitled 'Required System DFM (in the Full Study)' shows how the customer clarified the required processing using a Data Flow Diagram. However, the facility which the supplier proposes offers an interface to another system which the customer has not asked for. In this example it is an automatic link to the customer's Email system.

Another example of where an annotated version of the Required System DFM would be especially useful is where the proposed solution includes an application package offering functionality which lies outside the boundary of the requirement.

If there is insufficient time during the SPS for the supplier to prepare an annotated version of the whole of the DFM, the annotations should be limited to those processes where there is a mis-match between the IT support anticipated in the OR and the IT support being proposed by the supplier.

7.4 Evaluation of the SPS deliverables

The SSADM products developed by the supplier during the SPS are reviewed by the customer as part of evaluation of the SPS deliverables. The skills needed to carry out the review and the preparatory work required of the customer should not be underestimated.

At the time of evaluating the SPS deliverables, where these include application package software, the customer may benefit from the use of some of the SSADM specification products as a basis for evaluating the proposed packages. Guidance on this is provided in CCTA's Appraisal & Evaluation Library volume: *Overview and Procedures*.

When SSADM has been used during the SPS the customer needs a varied set of skills to fully review the SSADM products which the suppliers have developed. In order to carry out the review effectively, the customer needs:

- an understanding of the requirement which matches the level of detail in the suppliers' SSADM design products

- an understanding of the derivation of SSADM Stage 5 and Stage 6 products from the specification products of Stage 3

- an understanding of the supplier's approach to physical design and of the components of the suppliers' physical design, some of which may not have been documented using SSADM products.

It is outside the scope of this volume to provide guidance on the second and third of these skills. The remainder of this section, therefore, gives guidance on how the customer may acquire the first skill.

The most effective way of developing an understanding of the requirement at the level of detail needed to review a proposed physical design is for the customer to complete the Stage 5 activities, if this has not already been done during the Full Study. In some procurements this may result in the customer undertaking or completing Stage 5 activities in parallel with the SPS.

The benefits to the customer of preparing an 'in-house' set of Stage 5 products are:

- the customer develops a fuller understanding of the requirement and is, therefore, better able to answer detailed questions from the supplier

- the customer has a set of products which can be used to assess whether the supplier's equivalent products meet the requirement

- the customer is in a better position to make a fair assessment of the supplier's competence in the use of SSADM.

In practice, in order to achieve these benefits, it is not normally necessary for the customer to develop Stage 5 products which clarify all the functional requirements. For example, the customer may choose to develop only those process models necessary to understand the more complex events and enquiries. However, it may be that events and enquiries which seem straightforward at the time of preparing the OR are first shown to be complex by the SPS supplier. With regard to Dialogues, the customer may choose to develop only those for the most complex on-line functions.

If a supplier is not required under the SPS contract to develop process models, the customer-produced process models should be made available to the supplier, in order to help the supplier complete the SPS within the agreed timescale.

Where Stage 5 products are developed by the customer during the SPS, the customer may not insist that they form part of the requirement, because the products were not included in the OR as requirements. It is a matter for the customer and supplier to agree at the time of contract negotiation what the status of the customer Stage 5 products should be.

The development of Stage 5 products during the SPS may reveal that the wrong requirement has been specified (as distinct from a poorly specified requirement which the customer and the supplier interpret in different ways). In extreme cases, where the 'wrong' requirement includes mandatory requirements which need to be changed, it may be necessary for the customer to go back one or more stages in the procurement process, especially if the procurement is not under the 'Negotiated' Procedure of the EC Services Directive, since the short-list of suppliers may have been established on an unsound basis.

However, the cost to the customer of procuring a design which does not meet the real requirement is likely, in the long run, to be greater than the cost, in terms of slippage and additional expenditure, of reviewing the Requirements Specification, completing Stage 5 in full and re-commencing the procurement process.

Chapter 7
The System Procurement Study

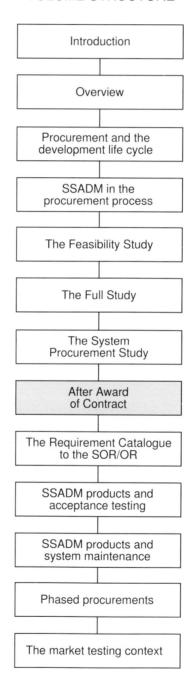

VOLUME STRUCTURE

8 After Award of Contract

8.1 Outline

This chapter addresses the use of SSADM products following the award of contract. Although EC/GATT regulations limit the ability to specify the use of SSADM at the SOR and OR stages, no such limitation exists at contract negotiation stage. A customer may thus seek, although they cannot impose, during contractual negotiations to specify that the supplier will use the SSADM design methods and procedures from Stages 5 and 6. In addition the customer is not inhibited from using SSADM products in activities carried out by his own personnel, such as acceptance criteria specification. This chapter provides some guidance on the products which may be of benefit to the customer after contract award and how they may be used to protect the customer's interests.

This chapter considers the potential use of SSADM products in the system life cycle activities after contract award, and their use in managing the supplier contract. These activities are as follows:

- monitoring system development
- Acceptance Trials
- Post Implementation Review
- systems support and maintenance.

They are discussed in the sections below.

Any products which are to be produced by the supplier to a particular standard or specification, including SSADM products, should have been specified in the contract.

8.2 Monitoring system development

The definitive acceptance of the supplied system should be based on the formal Acceptance Trials. However, the customer may well wish to monitor the supplier's progress during the development of the contracted system, rather than waiting for the system to be declared ready for acceptance on the due date.

If delays occur, or are likely to occur, in the schedule, then this could have a major impact on the customer's own business. In any case it will affect the training programme, system take-on and the activities of the user departments. Apart from any contractual considerations, the sooner that the customer is aware of any delays, either potential or actual, then the sooner it can start to plan for the repercussions and to limit any disruptions or other problems arising.

Conversely, if the supplier expects to be able to deliver the new system early, then the customer may be able to take advantage of this situation providing that his own plans are revised in sufficient time.

Some of the supplier's activities during system design and development could impact the customer's own responsibilities. They are likely to arise from the following:

- changes to the Requirements Specification, selected Technical System Option or other decisions agreed during the design and development process. These should always be subject to a Change Control procedure which has been agreed in the contract

- detailed design and implementation decisions which are made during the system design activities. These may affect the detailed operating procedures or the customer's planning. Early consideration of such factors will facilitate a smoother system implementation. The detailed specification of acceptance test data by the customer will probably be dependent, in part, on such detailed decisions (see Section 8.3).

If any changes or other decisions impact previous SSADM products, such as the Requirements Specification, then these products should be updated when the changes are confirmed. It is possible that the changes will have knock-on effects on other products, such as the Acceptance Criteria. Hence the customer must not only implement the primary changes arising but must also check rigorously for any secondary consequences.

In order that the customer may monitor the development effectively, it will agree a series of 'control points' with the supplier. The arrival at a control point should be demonstrable by the completion of one or more tangible products, such as a specification document. Each control point will have a planned date. The customer should be able to monitor progress by the timely completion of the relevant products by the respective planned dates.

The customer may wish to review some or all of these products on completion. This could also be in the supplier's interest since it would be more economic for the latter to identify and resolve problems earlier rather than later. However, these intermediate reviews must not prejudice the Acceptance Trials as the definitive arbiter of the system's acceptability.

If the supplier agrees to use SSADM as the design methodology, then the choice of control points may be based on the completion of significant SSADM products. For example, at the end of SSADM Step 540 Assemble Logical Design the logical design would be complete and this event could provide a suitable control point. Since the scope of the SSADM methodology finishes at the end of physical design, control points for the remainder of the project, where appropriate, would have to be based on other principles.

8.3 Acceptance Trials

The base requirements for acceptance testing are specified in the Acceptance Trials criteria which are derived directly from the Information Systems Requirements in the OR, as discussed in Chapter 10.

Acceptance testing is the means by which the customer verifies that the system meets the stated requirements. Within a procurement it is the customer's responsibility to perform acceptance testing; in fact it is the only form of testing for which the customer is responsible.

The contracted set of requirements may be different from those in the original Requirements Specification and the OR. Some of the requirements may have been non-mandatory and may have been omitted either from the supplier's proposal or during contractual negotiations. Hence the Requirements Specification will have been adjusted to those requirements for which there is a contracted solution.

Before Acceptance Trials can commence an Acceptance Trial Specification must be developed which contains the test data, test scripts and other details for the trials. The Acceptance Trial Specification will be based on the Acceptance Criteria in the contract and formally agreed between the customer and supplier. The detailed content of the Acceptance Trial Specification will also need to take account of the physical system design, for example, a test script may contain the detailed operational procedure to invoke a given system function.

Since the objective of the Acceptance Trials is to test whether the delivered system meets the contracted requirements, not the supplier's design, care must be taken to ensure that the physical design of a particular element is a valid implementation of the related requirements. Any discovered discrepancy in this matter – a requirement is not properly covered in the design or a design feature is an invalid interpretation of a requirement – should be notified to the supplier as soon as possible.

If the design has been carried out under the SSADM methodology, then the SSADM specification and design products may together be used in the development of the Acceptance Trial Specification. Chapter 10 describes how the SSADM specification and design products link together to provide requirements traceability paths which enable the implementation of individual requirements (and hence Acceptance Criteria) to be traced to the physical elements of the system, such as programs, database elements and hardware components. The test scripts and test data will need to take account of the design of these physical elements.

Acceptance testing is discussed in more detail in Chapter 10.

8.4 Post Implementation Review

It is the policy in Government that all IS projects should undergo a Post Implementation Review (PIR). The aims of a PIR are to determine whether the project has successfully achieved its objectives, including planned benefits, and to identify the lessons to be learnt from it. SSADM products can provide useful information for analyzing project results and potential problems arising.

The scope of a PIR covers many issues, including factors which are the supplier's responsibility, such as software performance, and those which are within the customer's domain, such as the administrative organization of the user staff. The list of planned benefits will be a significant element in the review and these will have been defined in the Feasibility Study report or Project Initiation Document and then refined during BSO selection in SSADM Stage 2.

The assessment of the supplier's performance will be derived largely from the results of the acceptance trials. Hence the SSADM products will not have a direct role (although they will have been used in deriving the test specifications, as described in Section 8.3).

In analyzing the project results and possibly the causes of problems or shortcomings, reference may be made to the SSADM products, as follows:

- versions of the Requirements Catalogue as they existed at the start of SSADM Stages 1, 2 and 3, respectively, and currently. This may help show if modifications to requirements were inconsistent with benefit expectations

- Current Services Description, which will help analyze the changes brought about by the new system

- versions of the User Catalogue and User Roles as they existed at the end of SSADM Stages 1 and 3, respectively, which will also help analyze the changes brought about by the new system

- the candidate Business System Options resulting from SSADM Step 210 as well as the option selected. This may also help to analyze the cause of any benefit shortfall

- the usefulness of the LDM, I/O Structures and events/enquiries as the basis for estimation using the Mk II Function Point approach. It may be necessary for the customer to calibrate its estimating method, that is, to review the industry average weights and coefficients in the light of the actual development. The ISE Library volume: *Estimating using Mk II Function Point Analysis* gives guidance on this issue.

8.5 Systems support

The drafting of the system development and implementation contract must take account of the maintenance of the system after it has gone into operation. The contract must stipulate either which design products will be maintained by the supplier in the course of system maintenance or which design products will be made available to the customer who may wish to take responsibility for system maintenance. The development and maintenance of the SSADM design products so that they may support system maintenance will increase the return on the investment in the SSADM Requirements Specification.

The evaluation of the supplier's proposals and the subsequent contractual negotiations will determine which SSADM design products should be developed, which should be used in system maintenance and, where appropriate, the form in which the design products will be transferred to the customer, for example, in the file format of a given CASE tool.

The system maintenance function will utilize both the SSADM specification and the SSADM design products. The former will relate the system to the business and user requirements. Thus the maintenance team can relate the system to the customer business, not just the context of the supplier's solution. The design products will demonstrate how individual requirements and the required functionality have been implemented. The tracing of requirements through to their physical implementation, as discussed in Section 8.3, is of benefit here.

The system maintenance function is discussed in more detail in Chapter 11.

VOLUME STRUCTURE

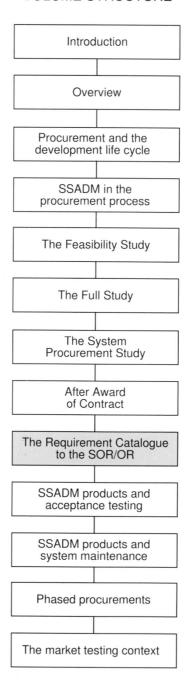

9 The Requirements Catalogue to the SOR/OR

9.1 Formatting the Requirements Catalogue

The Requirements Catalogue is the key SSADM product from which the information system requirements in the SOR and OR are derived. If the Requirements Catalogue entries are prepared with due and careful consideration to the procurement activities, then the subsequent effort in preparing the SOR and OR documents will be considerably reduced. The objective of this chapter is to ensure that the information system procurement requirements can be derived almost mechanically from the Requirements Catalogue with minimal editing while maintaining that product's essential role in the SSADM Requirements Specification.

The Requirements Catalogue is the central repository for information about the requirements. For the most part it is not sufficient by itself for the precise specification of the required system. Hence, a number of additional, more rigorous techniques and products are used to develop the requirements.

Some of the products which 'develop' the requirements are suitable as a means of amplifying requirements, that is, expressing the requirements to a lower level of detail. Other products are suitable as a means of clarifying the requirements, that is, providing additional information about the requirements without adding to the requirements. Some types of information system requirement, especially non-functional requirements, do not need to be specified in greater detail and are, therefore, specified only in the Requirements Catalogue.

There are compelling reasons why, on every project in which SSADM is used, the Requirements Catalogue should be the user's main point of reference:

- the scope of the Requirements Catalogue is the scope of the required information system

- the requirements are expressed in narrative format, thus making it easy for the user to understand

- many of the most important business issues in connection with the required information system are recorded in the Requirements Catalogue. Such as, who 'owns' a requirement, its benefit to the business and its priority.

Although the Requirements Catalogue may be the principal record of the (evolving) requirement initially and the first point of reference for the user, it has an equally important role to play in subsequent stages of the project. Chapter 6 describes the extent to which two of the major procurement products – the Statement of Requirement, and the Operational Requirement – should be made up of entries from the Requirements Catalogue. Chapter 10 describes the way in which the Requirements Catalogue may be used to 'drive' acceptance testing.

The purpose of this chapter is to describe how the Requirements Catalogue may be drafted so that it can serve all of these purposes.

The additional time and effort required to prepare the Requirements Catalogue in this way should be more than off-set by savings in the preparation of the SOR and OR, and in the planning stage for acceptance testing.

Adapting the Requirements Catalogue to the procurement products

It is an unnecessary duplication of effort to document the requirements in one format in the Requirements Catalogue and then rewrite the same requirements in another format in the relevant parts of the SOR and OR. On procurement projects, the Requirements Catalogue entries should be drafted from the outset in a form which enables the relevant parts of each entry to be 'pasted' into the SOR and the OR. However, the format of Requirements Catalogue entries, as described in the SSADM manuals, does not completely lend itself to this.

Most of the items on the SSADM Requirements Catalogue Entry form can be completed in the 'standard' way: source; owner; benefits; related documents; related requirements; and resolution. But for some of the items, principally the 'functional requirement' and 'non-functional requirement' items, a different approach should be taken. These differences are discussed below.

Identifier for each requirement

It is important to both the supplier and the customer that each requirement has its own identifier. In the proposal the supplier must describe, for each requirement, whether the requirement is met, how it is met, how well it is met and so on. In the evaluation exercise the easiest way for the customer to compare the proposals is by recording the suppliers' scores in some form of compliance table, in which each entry refers to one requirement and shows the scores awarded.

One requirement per identifier

SSADM does not prescribe the amount of information which should be recorded for functional and non-functional requirements. It is, therefore, acceptable SSADM practice for an entry in the Requirements Catalogue to consist of many paragraphs which describe, for example, functional requirements; each paragraph may contain many specific requirements. Hence in some cases many requirements may only be identified by reference to a common Requirements Catalogue entry number. However, on a procurement project, each requirement identifier must identify one requirement and one requirement only.

'Normal' SSADM Requirements Catalogue requirement description	Comment on 'normal' SSADM entry	Requirement description suitable for 'pasting' into SOR/OR
Reqt. 100 Communications Standards The communications on the current network are all based upon X.400(88), and the future system must use this standard, both for communications within Security Division, and between the Security system and other systems on the network. The modems must comply with BS 6320, and must comply to the appropriate level of CCITT V.	This 'requirement' is actually three requirements, each of which must be stated separately and must have its own identifier.	*Reqt. 100 Communications Standards* 100a. The message format for communications within the required system and between the required system and other systems must be X.400(88). 100b. Modems must be BS 6320 compliant. 100c. Modems must comply to the appropriate level of CCITT V.

Table 9.1: Conversion of complex requirement to multiple elementary requirements

Table 9.1 shows how one complex requirement entry should be divided into several discrete requirements to facilitate their transfer to the procurement documents.

To continue this argument, SSADM provides for a functional requirement to be grouped in the same Requirements Catalogue entry as the non-functional requirements which are specific to the functional requirement. These non-functional requirements could cover several distinct topics, such as response times and access restrictions. In these circumstances it could be appropriate to consider the functional requirement and each of the non-functional requirements, respectively, as separate procurement requirements. An example is given in Table 9.2.

'Normal' SSADM Requirements Catalogue requirement description	Comment on 'normal' SSADM entry	Requirement description suitable for 'pasting' into SOR/OR
Reqt. 203 Customer account creation *Functional requirement* A system operative must be able to create a new user account. *Non-functional requirements* 1: The response time for creating a new user account must be less than 6 seconds. 2: The response time for creating a user account should be less than 3 seconds for 90% of all transactions. 3: This function must only be available to personnel authorized by the System Manager.	The requirement description covers a number of distinct, albeit related, requirements. For procurement purposes they should be separated as shown on the right.	*Reqt. 203 Customer account creation* 203a. A system operative must be able to create a new user account. 203b. The response time for creating a new user account must be less than 6 seconds. 203c. The response time for creating a user account should be less than 3 seconds for 90% of all transactions. 203d. The customer account creation function must only be available to personnel authorized by the System Manager.

Table 9.2: Conversion of functional and non-functional requirements to distinct procurement requirements

Where a requirement cannot be fully understood by the supplier without some explanation, the explanation should be given in a paragraph of the requirement which has no identifier and which contains no 'hidden' requirement. The explanatory comment can be highlighted in some way, possibly by enclosing it in braces ({}). This emphasizes its purpose to the supplier and facilitates its removal when the OR document is converted to acceptance criteria at a later point in time. An example is given below.

Reqt. 84 Help Desk
{The envisaged purpose of the help desk is to ensure that the supplier accepts ownership of a system error or fault.}

84a. The supplier must provide a help desk 24 hours a day, 365 days a year.

84b. The help desk must initiate action to rectify a problem within 2 hours of being notified of an error/fault report.

Example use of braces to distinguish comment text from actual requirements

Measurability of each requirement

Each requirement must be described in a manner that allows the solutions offered by suppliers to be tested. Hence **the wording of the requirement alone** should be the yardstick on whether a supplier's proposed solution complies with that specific requirement. For functional requirements, it must be possible to determine whether a supplier's proposed solution includes the specified functionality or not. For data requirements, including those amplified in the Logical Data Model, it must be possible to determine whether a proposed solution includes support for the data requirement or not. For non-functional requirements, it must be possible to determine whether the supplier's solution meets the minimum acceptable quality.

A supplier's solution will be tested against the requirements at all stages of the procurement process from the receipt of proposals until either the solution is discarded or the procured system is accepted. These requirements will form the basis for evaluating the proposals, technical discussions with the suppliers, drafting the relevant contract schedules and preparing the system acceptance tests.

The following items in the Requirements Catalogue Entry form in the SSADM manuals should not be used:

- non-functional requirement target value

- non-functional requirement acceptable range.

The target value and acceptable range information is still documented but not in the way suggested in the SSADM manuals. If a non-functional requirement has both a target value and an acceptable range, the requirement must be stated twice. One version expresses the target value as a desirable requirement. The other version has the 'worst' figure of the acceptable range as the least acceptable level for compliance with the requirement. In the latter case, if the related functional requirement is mandatory, then this non-functional requirement is also mandatory.

Table 9.3 shows how this principle can be applied to a 'normal' SSADM Requirements Catalogue entry.

Business objectives are requirements

The business objectives which the information system procurement is addressing should be included in the Requirements Catalogue. Business objectives include not only actual business objectives from the organization plans or from the IS Strategy but also business requirements and performance improvement measures identified in the Feasibility Report.

In the analysis and specification stages the presence of the business objectives in the Requirements Catalogue helps users and analysts to remain focused on the key areas of the requirement. The business objectives should be 'pasted' into the Background to the Requirement section of the SOR and OR since they provide useful guidance to the suppliers.

'Normal' SSADM Requirements Catalogue entry	Comment on 'normal' SSADM entry	Requirement description suitable for 'pasting' into SOR/OR
Reqt. 111 Passenger Transfers *Priority* = Mandatory *Functional requirement* The system must enable the Booking Clerk to transfer all passengers from a cancelled flight to another flight without having to create new records for the passengers. *Non-functional requirement* Operating procedure *Target* = automatic transfer of all passenger records. *Acceptable range* = single key depression for each passenger transfer.	To suit the SOR or OR this requirement should be stated twice: once as a mandatory requirement measured by means of the acceptable range, and once as a desirable requirement measured by means of the target value.	*Reqt. 111 Passenger Transfers* 111a. The system must enable the Booking Clerk to transfer a passenger record from a cancelled flight to another flight by means of a single key depression. [MANDATORY] 111b. The system should be capable of automatically transferring all of the passenger records from a cancelled flight to another flight. [DESIRABLE]

Table 9.3: Example of how non-functional requirements can be expressed in procurement terms

The Requirements Catalogue entries which state the business objectives are 'resolved' only by means of other Requirements Catalogue entries and not by means of other SSADM techniques and products.

Since they do not form part of the requirement in the procurement sense, entries in the Requirements Catalogue which record business objectives need not be expressed in the style which applies to the information system requirements, eg the business objectives need not be expressed in terms which are measurable.

A useful check on the priority of a requirement is to insert in the catalogue a cross-reference from the requirement to the business objective(s) which it addresses. Not all requirements address business objectives directly. However, the discipline of testing which business objective is being addressed helps to identify requirements which are not essential or which, in extreme cases, are outside the scope of the project.

Design constraints are requirements

The design constraints which apply to the proposed system should be included in the Requirements Catalogue. Design constraints can relate to almost any aspect of the requirement, but the three most common types of constraint are:

- standards (such as software, communications and safety)

- regulations (such as those under the Health and Safety at Work Act)

- interfaces to other systems.

In the analysis and specification stages the presence of the constraints in the Requirements Catalogue ensures that their impact on the remainder of the requirements will not be overlooked. Some of these design constraints may be resolved by means of other SSADM products, others may not. For example, a data interface constraint should be resolved by the I/O Structure which defines the interface in detail, whereas the constraint of meeting certain requirements of the Health and Safety Regulations would not be resolved by an SSADM product, but by the nature of the goods offered by the supplier.

Where a design constraint requirement is resolved by some other requirement (such as a functional requirement), then it should not be included explicitly in the SOR and OR. If it is not resolved elsewhere, the design constraint requirement should be 'pasted' into the Design Constraints section of the SOR/OR. If an original design constraint is only partially resolved by other requirements, then a new requirement should be formulated to cover just the unresolved part of the constraint and this should be 'pasted' into the Design Constraints section of the SOR/OR.

Requirements Catalogue in sections

The procurement products, the SOR and OR, have pre-defined sections, which are intended to help suppliers understand the requirement. In order to facilitate the incorporation of Requirements Catalogue entries directly into the procurement products, the Requirements Catalogue should be divided into the same sections.

The Requirement Catalogue should be divided into sections as follows:

- functional requirements, grouped by functional areas. Special non-functional requirements (such as access requirements which are peculiar to a particular function) should be grouped with the related functional requirements

- system characteristics, including non-functional requirements which are system-wide or apply to large groups of functions

- system operations

- design constraints

- service requirements, such as user support, maintenance, training and documentation.

As indicated in Section 9.1 the Requirements Catalogue is not the only product required to completely specify the required system. Further detail may be expressed in other SSADM products, such as the Data Flow Model and the Logical Data Model. These other SSADM products will normally be published as annexes to the SOR and OR and will be cross-referenced from the Requirement Catalogue entries. During SSADM Stages 1 to 3, these cross-references will just use the standard SSADM identifiers. While these identifiers will continue in use within the SOR/OR, it would be useful to add further cross-reference information, such as annex identifiers or page numbers.

The mere presence of an annex in the SOR/OR does not itself imply that its contents are part of the requirement.

The relevance of particular parts of an annex must be specifically stated in a requirement, as shown below.

> Reqt. 204 Customer data
> The system must be capable of holding data on customers. As a minimum this must comprise the data structure and attributes specified as the CUSTOMER entity in Annex D.

Example cross-reference to another SSADM product (the Logical Data Model)

9.2 Level of requirement detail

Requirements need to be comprehensively described and precisely identified. They must also be expressed in a manner that allows them to be used to test the solutions offered by suppliers. Compliance with the wording of a given requirement should ensure a high confidence that that aspect of the customer's needs would be adequately met by the proposed solution. However, the wording should neither attempt to predict the solution to the requirement nor constrain the potential solution with unnecessary or over-prescriptive detail.

It is important that the requirements specify what the customer **requires** and **not a potential means of implementing** that requirement. Otherwise there is a high risk that some potential solutions to the customer's real needs, possibly better solutions than that previously envisaged by the customer, will be disqualified unnecessarily.

Where the solution to a requirement is likely to be met by an application package, the level of detail should be low enough to determine whether a potential package can fully meet the requirement but high enough to avoid constraining the way in which the requirement can reasonably be met.

Mandatory requirements must not contain any non-essential detail. This is because failure to comply with a single mandatory requirement must exclude the supplier's proposal from further consideration unless the procurement is under the 'Negotiated' Procedures of the EC Services Directive, where some flexibility on this point may be possible. An illustration of this point is given in Table 9.4.

Original Requirement	Comment on Original Requirement	Revised Requirement
The system must provide a report of journeys made, including: journey date, journey length, start point, end point, journey time, and carrier. [MANDATORY]	Some of these items such as, the journey time, are not essential to the user.	The system must provide a report of journeys made. This report must include as a minimum: journey date, start date and end date. [MANDATORY] The report of journeys made should also include: journey length, start point, end point, journey time and carrier. [DESIRABLE]

Table 9.4: Reduction of detail in mandatory requirements

The revised version of the above requirement allows suppliers to propose a more generic report. Hence the customer may avoid incurring costs from suppliers who have to extend their package to include the additional data items.

Chapter 6 discusses the circumstances in which other SSADM products may be included in the SOR and OR to amplify Requirements Catalogue entries.

9.3 Defining mandatory and desirable requirements

The mandatory and desirable priorities must be allocated in time for BSO selection, or for selection from among Feasibility Options if the procurement is to be initiated upon completion of the Feasibility Study. A supplier's proposal must be rejected if it fails to comply with **any** mandatory requirement. Hence, before classifying a particular requirement as mandatory, the customer must be sure that it is the intention to reject a proposal merely for not meeting that requirement alone. As previously stated, if the procurement is under the 'Negotiated' Procedures of the EC Services Directive, some flexibility on compliance with all mandatory requirements may be possible.

SSADM does not prescribe the regime which should be used for prioritizing requirements. Thus it is acceptable SSADM practice in non-procurement Full Studies to allocate priorities such as high, medium, low, mandatory, optional, desirable, et alia. However, if requirements are to be 'pasted' into an SOR or OR, then the priorities must be 'mandatory' and 'desirable'.

Where the customer wishes to give guidance to suppliers on the relative importance of desirable requirements, the requirements may be annotated 'high', 'medium' and 'low'. The additional annotations indicate to the suppliers that those requirements are allocated greater and lesser weightings in the evaluation model. An alternative approach is to replace the single 'desirable' priority with 'highly desirable' and 'desirable', respectively.

Mandatory requirements exist as a single class of priority; one may not give mandatory requirements a sub-priority.

In order to assist both the supplier and the customer's evaluation team, each requirement in the OR (and the SOR if it contains non-mandatory requirements) should be annotated with its priority. This is illustrated by the '[MANDATORY]' and '[DESIRABLE]' labels in the earlier examples in this chapter.

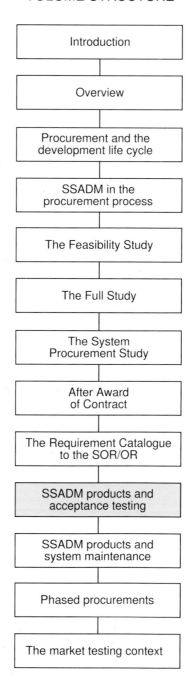

10 SSADM products and acceptance testing

Acceptance testing is the means by which the customer verifies that the system meets the stated requirements. Within a procurement acceptance testing is the responsibility of the customer. Acceptance testing is conducted against Acceptance Trials criteria and in accordance with Acceptance Trials procedures, both of which are contained in the schedules to the implementation contract. Many of the Acceptance Trials criteria can be derived from SSADM products.

Before acceptance testing can commence, an Acceptance Trial Specification is drawn up. This contains test scripts and other details of the trials. If an SSADM design has been agreed as part of the procurement, the quality of the Acceptance Trial Specification can benefit from requirements traceability paths, which trace from the requirements through the design to the physical components of the implemented system.

Acceptance testing is the only type of testing for which the customer is responsible. Responsibility for the other forms of testing, unit, integration and system testing, lies primarily with the supplier, although the customer may wish to be a part of them or, at least, scrutinize the results. If the customer is satisfied that requirements have already been proven by supplier testing, it may be possible for the customer to simplify some of the acceptance tests. However, any simplification must be considered with a full acceptance of the increased risk to the procurement.

10.1 The role of the Acceptance Trial

Acceptance of the delivered system is conditional on the successful completion of one or more Acceptance Trials. One of the benefits to the customer of using SSADM during the Full Study is that the Acceptance Trials criteria, against which the acceptance tests are made, can be derived directly from the SSADM specification products.

Following each Acceptance Trial the customer analyses the results, and decides whether to accept, conditionally accept or fail the trial. Conditional acceptance can

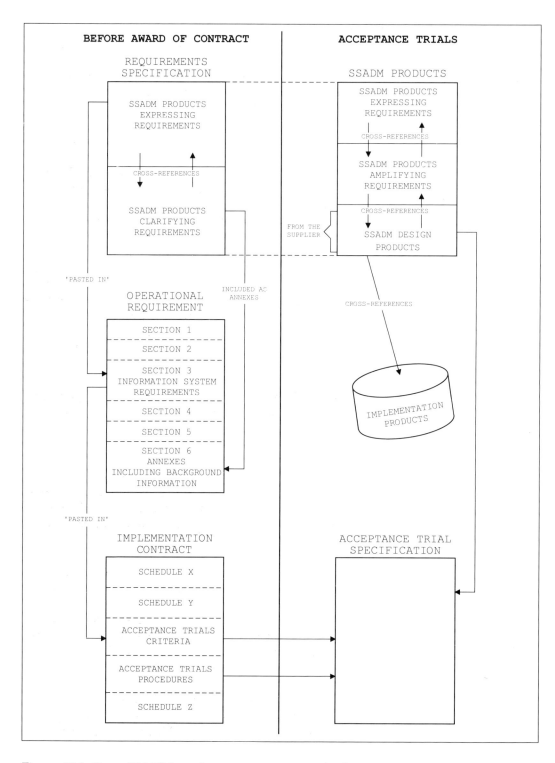

Figure 10.1: From SSADM products to acceptance criteria

occur when the principal criteria are met but minor problems are left outstanding at the end of the trial.

In the case of acceptance, it is normal for the title to the system to transfer to the customer, the customer pays for the system, and live operation can commence.

It is of paramount importance that the criteria for the trials are based on the customer's requirements as specified in the Operational Requirement (OR). This is essential if the customer is to ensure that the procured system achieves the value-for-money targets planned, and specified in the Business Case for the project. It is also the normal contractual condition.

The use of SSADM in the Full Study ensures that the OR contains as many requirements as possible taken directly from the Requirements Specification. Chapter 6 describes how, in a procurement Full Study in which SSADM is used, the functional, data and non-functional requirements in the OR are 'pasted in' from SSADM products.

Figure 10.1 shows the route by which SSADM products from the Requirements Specification find their way into the acceptance products.

The arrows in the diagram, in the column headed 'BEFORE AWARD OF CONTRACT', which join the Requirements Specification to the OR indicate the way in which SSADM products are copied into the OR, either to express requirements or as background information to clarify the requirements.

The arrow in the diagram which joins the OR to the implementation contract indicates the way in which Acceptance Trials criteria are copied from the requirements in the OR (the diagram is concerned only with the use of SSADM products, and does not indicate the extent to which other sections of the OR are copied into the Acceptance Trials criteria).

Thus, some, but not all, of the SSADM products of the Full Study find their way, by the time of award of contract, into the product by which the customer must evaluate the delivered system. This helps to ensure that the Acceptance Trials will test whether what the supplier has provided is what the customer originally asked for.

However, it does not tell the customer how to compile test scripts. SSADM can be of use to the customer in the preparation of test scripts by means of the requirements traceability which it provides.

Figure 10.1 indicates the traceability paths, in a simplified form, in the Requirements Specification box in the 'BEFORE AWARD OF CONTRACT' column. This box is split into two. One half represents the SSADM products which express requirements; the other half represents the SSADM products which clarify the requirements. The arrows which cross between the two halves represent the cross-references between the SSADM products. These cross-references provide the start of the requirements traceability paths. The second column in the diagram contains three major products of the project at the time of the Acceptance Trials. The first box in the column represents the SSADM products. The diagram assumes that by the time of the Acceptance Trials, the supplier has developed an SSADM design. The SSADM design products should cross-refer back to the specification products from which they are derived; they should also cross-refer to the physical products by which the system has been implemented. In this way the customer's and the supplier's SSADM products in combination provide full requirements traceability paths.

The implications of requirements traceability, and the SSADM products which provide the traceability paths, are described in the following section.

10.2 SSADM traceability paths

An important characteristic of an information system for which IT support has been developed is the requirements traceability within the system. It must be possible to tell from the documentation of the system how the requirements have been met by the IT implementation. The SSADM specification products and the SSADM design products taken together provide full requirements traceability.

This section identifies which SSADM specification products must be annotated by the customer, together with the cross-references needed to support the traceability paths.

It also identifies the SSADM design products which must be annotated by the supplier, together with the cross-references needed to complete the traceability paths.

10.2.1 Benefits of requirements traceability

The requirements traceability paths provide a powerful management tool with which to undertake:

- preparation for acceptance tests
- maintenance of the operational system.

If the supplier extends the traceability paths from the customer's specification products into the SSADM design products, the customer benefits from being able to tell which implemented components deliver the requirements. This information is useful in the preparation of acceptance test scripts.

The advantage of having this information available may be appreciated by considering the position when requirements traceability is not provided. At the time of preparing test scripts when the requirements traceability is not provided, the customer is obliged to refer to the supplier's physical design, in order to identify the functions to be tested. The design may document the physical functionality very well, but if the design does not cross-refer back to the requirements, it is hard for the customer to be sure which physical processes relate to which functions in the Requirements Specification. The problem may be graphically summarized – see Figure 10.2.

Nor can the customer easily compile test scripts by approaching the problem from the other direction, that is, by taking each criteria and identifying all of the functionality to be tested, especially since the design is likely to be expressed in physical terms.

In both cases the risk to the customer is that the Acceptance Trials will test whether what has been supplied is useful, rather than whether what has been supplied is what was contracted.

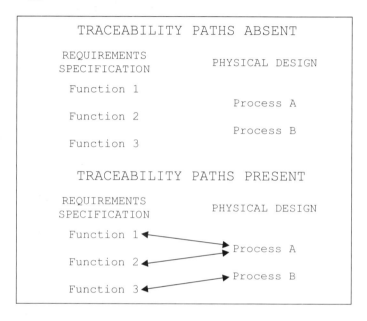

Figure 10.2

The benefit of the requirements traceability to the customer in the longer term is that maintenance of the system can be based on documentation which links requirements to the physical system being maintained, and vice versa.

Chapter 11 describes how SSADM products can be used to maintain the system. The remainder of this chapter describes how SSADM products can be used to prepare for acceptance testing.

10.2.2 SSADM products providing traceability

Many of the cross-references between products needed to put the requirements traceability paths in place are described in the SSADM manuals. Therefore, much of the work of putting the paths in place is done on any project on which SSADM is used. The time and effort required to complete the paths by inserting additional cross-references should be more than offset by savings in the acceptance testing and maintenance activities.

Figure 10.3 shows the products and the cross-references between them which provide the requirements traceability paths in SSADM. They are presented in four columns:

- the Requirements Catalogue, the central repository for all the requirements

- the SSADM specification products

- the SSADM design products

- the implemented products of the operational system.

The paths, ie cross-references, provided by SSADM products are shown on the diagram as solid arrows; the perforated arrows indicate cross-references which have to be provided by non-SSADM products.

SSADM finishes at the physical design stage. It is, therefore, possible to cross-refer from the SSADM design products to the physical products which implement them, but cross-references in the opposite direction must be provided using non-SSADM products.

The balance of this section addresses each column of Figure 10.3 in turn, describing in more detail each of the boxes in the figure and their relationship.

Section 10.2.3 discusses how software tools can be used to implement the traceability path between the design products and the physically implemented products.

Column 1 – Requirements Catalogue

For the purposes of illustrating the requirements traceability paths, the figure includes some, but by no means all, of the major categories of requirement:

- data requirements

- functional requirements (the features and facilities required)

- non-functional requirements applying to functions (how well the functions are required to perform, ie to what quality level)

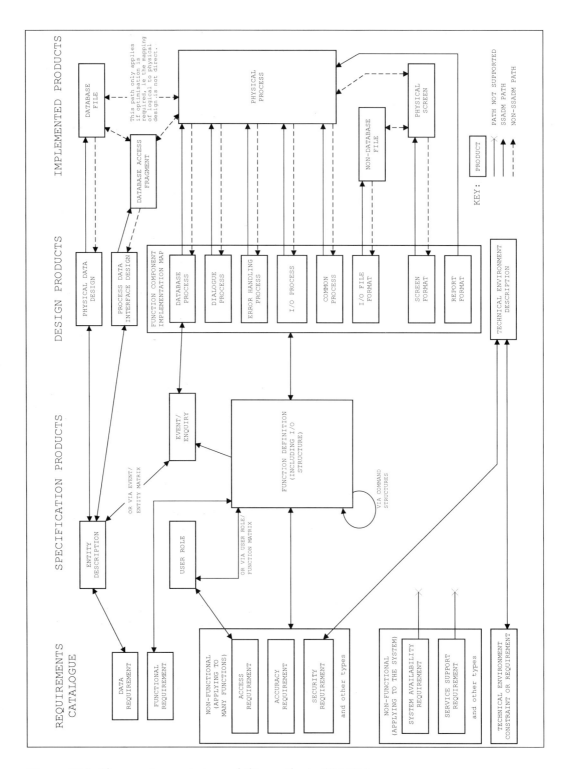

Figure 10.3: The requirements traceability paths in SSADM

- non-functional requirements applying to the system as a whole, eg system availability requirements

- technical environment constraints/requirements.

For all the above types of requirement the requirements traceability paths begin with cross-references from entries in the Requirements Catalogue to the SSADM products which amplify or clarify the requirement. These cross-references are made in the 'resolution' field of entries in the Requirements Catalogue.

Data requirement entries

Data requirements should cross-refer to the relevant Entity Descriptions. Here is an example of a data requirement:

In the current system Form 242 Equipment Tracking (copy attached) is used to record items of security equipment and the buildings in which they have been installed. The future system must be capable of holding these details.

This requirement should cross-refer to the Entity Descriptions for the entities SECURITY EQUIPMENT ITEM and BUILDING, where the detail about the data requirement is expressed.

The traceability path can be simplified, if data relationship requirements are documented in the Entity Descriptions. This removes the need to maintain Relationship Descriptions in the traceability paths. For example, consider this data requirement:

The system must hold details of the last 3 buildings in which the security equipment has been installed.

In this example, reference to the one-to-three relationship should be made in the appropriate Entity Description.

The clarity of the requirements traceability paths is diminished if data and functional requirements are expressed together as a single requirement. For example, consider this requirement:

The system must hold details of security equipment, and must provide the Security Manager with a quarterly statement of the current location of each item.

In this example, the data and functional requirements should be separated. Chapter 9 gives guidance on converting complex requirements to multiple elementary requirements.

It is essential that each requirement, ie each entry in the Requirements Catalogue, has a unique identifier, so that cross-reference may be made to it in other products. With its own identifier, every requirement can have a unique position at the start of the traceability paths.

In the example quoted above, once the data and functional requirements have been separated and each given a unique identifier, the data requirement can be 'resolved' by means of Entity Descriptions, and the functional requirement can be 'resolved' by means of Function Definitions. Two paths are available. One from the data requirement, the other from the functional requirement.

Functional requirement entries

Functional requirements should cross-refer to all of the relevant functions. Consider this example of a functional requirement:

The system must enable users to maintain an index of security equipment types.

This requirement should cross-refer to the Function Definition which gives details about the index maintenance function.

Figure 10.3 assumes a fairly stable user organisation, in practice the link between Functional Requirement and Function Definition may not be direct. Function Definitions may be established via prototyping (by re-packaging events/enquiries for different user roles).

As for data requirement entries, a functional requirement entry should not refer to two or more functional requirements, or include data requirements. For example, consider this functional requirement:

The system must enable users to maintain an index of equipment types, and must be capable of reporting upon all changes to the index in a twelve-month period.

In this example, the functional requirements should be separated to become two entries: one which specifies the maintenance requirement; another which specifies the enquiry requirement.

In SSADM an entry in the Requirements Catalogue which describes a functional requirement may also describe non-functional requirements which apply specifically to that functional requirement. These non-functional requirements form part of the acceptance criteria for the functional requirement.

Non-functional requirement entries (applying to many functions)

Figure 10.3 refers to only three types of non-functional requirement entries which apply to many functions:

- access requirement entries

- accuracy requirement entries

- security requirement entries.

Other types of non-functional requirements which apply to (many or all) functions are not shown.

All such entries should, where appropriate, cross-refer to all of the functions to which they apply. Consider this example of such a non-functional requirement:

When calculating customer discount levels, the system must round down to the average of all of the customer's accounts.

In this example, many functions may include the calculation of customer discount levels. The requirement should cross-refer to all of the relevant Function Definitions.

The two other categories of non-functional requirement entries which apply to many functions are shown explicitly on the diagram, because they cross-refer to other products than the Function Definitions.

Firstly, any access requirement entries which specify precisely who may access the system should cross-refer to the User Roles. Consider the following requirement:

The system must restrict access to update functions to staff in the Security Division.

In this example, the entry in the Requirements Catalogue should cross-refer to the user roles identified for staff in the Security Division. Access requirements which do not specify precisely who may access the system (for example, 'the system must prevent unauthorized users from accessing the centrally held files'), are not cross-referred to the User Roles.

Secondly, many security requirements are 'resolved' without further specification. Wherever possible such security requirements should cross-refer to the parts of the Technical Environment Description, which describe the physical nature of the solution. Examples of this type of requirement are:

- physical security requirements
- back-up requirements
- off-site storage requirements
- fall-back requirements
- contingency requirements.

Non-functional requirement entries (applying to the system)

Figure 10.3 refers explicitly to only two types, there are others, of non-functional requirements applying to the system:

- system availability requirements
- service support requirements.

SSADM products provide no requirement traceability paths from these types of requirement.

Technical environment constraints and requirements	The name in the final box in the first column of Figure 10.3 refers to all of the requirements which are not specified in further detail in any SSADM product, and which are 'resolved' in ways described in the Technical Environment Description (TED). Such requirements should cross-refer to the appropriate part of the TED. Examples of this type of requirement are: • hardware requirements • software requirements • network and telecommunication requirements • hardware/software/network constraints • take-on requirements.

Column 2 – SSADM Specification Products

Entity Descriptions	Entity Descriptions may either amplify data requirements, ie express data requirements in more detail, or they may clarify data requirements, or they may do both. Each Entity Description should cross-refer to all of the data requirements to which it relates. The traceability path from entities to the events and enquiries which access and update them is provided by means of: • Entity Life Histories (ELHs) • Effect Correspondence Diagrams (ECDs) and/or Update Process Models (UPMs) • Enquiry Access Paths (EAPs) and/or Enquiry Process Models (EPMs). The traceability path from the entities to the events and enquiries is summarized by the Event Entity Matrix (EEM). The EEM provides a summary which is more helpful if it has been extended to include enquiries as well as events, and 'read only' actions as well as 'create',

'modify', and 'delete' – as suggested for the maintenance environment in Chapter 11.

When the Physical Data Design has been agreed in Stage 6, cross-reference should be made on each Entity Description to the physical entity, or entities, which correspond to the logical entity.

If there is a mismatch between the way(s) in which an entity is logically accessed and the way in which it is physically accessed, then the Entity Description should cross-refer to the part of the PDI design which describes how the entity is accessed physically.

User Roles

Each entry in the User Roles should cross-refer to all of the access requirements which it clarifies – see 'non-functional requirement entries (applying to many functions)' above.

The cross-reference from each user role to the on-line functions available to the user role is provided by means of the User Role/Function Matrix.

Function Definitions

Each Function Definition should cross-refer to:

- the functional requirement(s) which it 'resolves', ie which it clarifies

- the non-functional requirements which apply to the function.

It is important that cross-reference is made to all of the non-functional requirements which apply to the function, since these specify quality criteria by which the function's acceptability is assessed.

When SSADM is used in a procurement Full Study, the service level requirements of the Function Definition should not be used, for the reasons given in Chapter 6. (The function-specific service level requirements should be recorded in the Requirements Catalogue as non-functional requirement entries which apply to the relevant functional requirement entry.)

For the purposes of acceptance testing it is important to know which functions are closely related to one another. For example, for an off-line function there may be a requirement for it to be performed in the same run-time batch flow as other off-line functions. This cross-reference must be recorded on the Function Definition. For an on-line function it is likely that the Command Structure provides the relevant cross-references. If not, they should be included in the Function Definition.

It is important to know which events and enquiries are included in the function. Therefore, each Function Definition should include appropriate cross-references. This part of the requirements traceability path operates in one direction only, ie from function to event (hence the one-way arrow on the diagram). The event to function path is provided by means of the Function Component Implementation Map (FCIM).

Events and enquiries   Events and enquiries are an essential part of the traceability paths, because in SSADM, data requirements are linked to functional requirements at the event/enquiry level rather than at the function level. However, there are no event and enquiry descriptions in SSADM. This leads to a certain amount of complexity in the requirements traceability paths.

Events and enquiries are referred to in the Function Definitions, but none of the products which describe the events and enquiries, ELHs, ECDs/UPMs, EAPs/EPMs or the EEM, cross-refer to functions. This is why the arrow on the diagram between Function Definitions and Events/Enquiries is one-way only. In SSADM the path from events and enquiries to functions is documented only in the FCIM (see below).

Column 3 – SSADM Design Products

Physical Data Design   In procurements the development of the Physical Data Design is the responsibility of the supplier. The description of an entity in the Physical Data Design should cross-refer to the logical entity, or entities, to which the physical entity corresponds. It should also cross-refer to the name of the database file or table which implements the physical entity. In this way, it becomes possible to trace which physical file(s) implement the entities in the Logical Data Model.

Process Data Interface (PDI) design	SSADM does not prescribe the format of a PDI design. However, it is likely to consist of sections, or equivalents, each of which addresses one mismatch between the way an entity in the LDM is accessed and the way it is physically accessed. Each section, or equivalent, in the PDI design should cross-refer to the logical entity which is the subject of the mismatch. It should also cross-refer to the name of the database access fragment which implements the physical access of the entity.
Function Component Implementation Map (FCIM)	In procurements the development of the FCIM is the responsibility of the supplier. The method does not prescribe the format of an FCIM, but its purpose is to specify the components of a function not specified elsewhere, and to link the logical components of functions to the physical objects which implement them.

The FCIM is compiled in parts which map one-to-one with a Function Definition. The FCIM contains the specifications and the designs, as appropriate, for the following components of each function:

- database processes
- dialogue processes
- error handling processes
- I/O processes
- common processes
- I/O file formats
- Screen Formats
- Report Formats.

The FCIM indicates the re-useability of these function components, ie it shows all of the functions in which each component is used.

The database processes must cross-refer to the events or enquiries which trigger them, since the FCIM is the only SSADM product which supports traceability paths from events/enquiries to functions.

Each component in the FCIM should also cross-refer to the name of the physical product which implements it. In this way it becomes possible to trace:

- the physical process(es) which implement the functions, events and enquiries

- the physical non-database files which implement the functions' input and output files

- the physical screens which implement the Screen Formats.

Technical Environment Description (TED)

In procurements the development of the TED is normally the responsibility of the supplier, though in certain circumstances it may be appropriate for the customer to prepare an initial version of the TED during the Full Study – see Chapter 6 for details.

The format of a TED is not prescribed in SSADM. The TED defines in detail many of the functional and physical aspects of the system which are essential to the day-to-day running of the system – whether during acceptance testing or when the system has gone into operation – such as:

- the system hardware

- the system software

- the system network and telecommunications hardware and software

- fall-back and recovery arrangements

- access and security arrangements

- hardware/software maintenance

- system sizing.

Wherever appropriate, references should be made in the TED to the requirements from which the components of the system are derived. In this way, it becomes possible to check not only that the physical system contains all of the components which were required, but also, vice versa, to identify which components of the system support the requirements, and which have become a part of the system for other reasons!

Column 4 – Implemented Products

Figure 10.3 summarizes a typical set of (software) products which implement IT support for an information system:

- database files

- database access fragments (from SSADM designs only)

- physical processes

- non-database files

- physical screens.

It is possible to cross-refer from the SSADM design products to the physical implementation products which implement them, but cross-references in the opposite direction, and cross-references between implemented products, must be provided using non-SSADM products, such as a Data Dictionary System (DDS). These paths are shown on the diagram as perforated arrows.

10.2.3 Software tools support

The technical environment in which acceptance testing of the system is undertaken may include sophisticated software tools, such as a DDS, provided by the supplier from his development environment. If it does, then the tool can be used to automate cross-references from the design products to the implemented products, and vice versa.

If the software tool is able to hold design products, such as an FCIM, then the customer and supplier should agree which traceability paths will be documented in the SSADM design products (by the supplier), and which will be documented using the software tool. In reaching this agreement, the customer should take into account the likelihood of a third party maintaining the system in the future, and whether the software tool would be made available to the third party.

10.3 SSADM products in acceptance testing

SSADM products are used during the Full Study to document acceptance criteria against which a delivered system may be measured. However, SSADM does not include a task list for performing acceptance testing. This section gives guidance on how to use SSADM products in preparing for Acceptance Trials.

There are significant advantages to the customer of basing acceptance tests on SSADM products to the greatest extent possible:

- the customer can work from products which are familiar rather than from products of the supplier which are unfamiliar or which cannot be understood without training

- tests are made against Acceptance Trials criteria derived from the SSADM Requirements Specification, rather than against acceptance criteria drawn up during development which may be influenced by the strengths and weaknesses of the supplier's approach to design and build, eg the strengths of the supplier's development software toolset

- the time and effort to plan the acceptance tests is greatly reduced, because for the most part the test paths, and the products required to test each path, are already documented in the traceability paths of SSADM.

During the Full Study the customer uses some SSADM products to express requirements, and other SSADM products to clarify the requirements – Chapter 6 gives guidance. The SSADM products which express requirements are included in the implementation contract schedules as Acceptance Trials criteria. These SSADM products, in conjunction with the products which clarify the requirements and with the design products developed by the supplier, document the requirements traceability paths. Thus, for the acceptance tests to which the SSADM products relate, the SSADM products provide two sorts of information needed by the customer. They describe what the measurement is by which the test is passed or failed, and they provide all the detail about the requirement which the customer needs to compile the test script.

The extent to which SSADM products are useful in the preparation of test scripts depends on whether the test to be scripted is to test:

- functional requirements, including the non-functional requirements which apply to them, also including the relevant data requirements

- non-functional requirements which apply to the system (as opposed to its functions).

10.3.1 Testing functional requirements

The Function Definitions and the Entity Descriptions are the core products within the Requirements Specification for deriving the test criteria. The requirements which the Function Definitions and Entity Descriptions satisfy can be identified by examining the requirements traceability paths. It is usually the case that many non-functional requirements apply to each function.

The IT supported processes which the customer needs to access for the test can be traced by means of the function's entry in the FCIM.

After all the relevant acceptance criteria and computer processes have been identified, the customer prepares two types of test for the function:

- business functionality tests

- performance tests.

Business functionality tests	This form of test is concerned with ensuring that the needs of the business will be met, as required in the functional requirement and as clarified in the Function Definition.
	Each function is considered as a unit that must be tested. To achieve this, acceptance criteria and expected test results are defined to test individual components of a function.
	The components of a function are:
	- I/O component
	- dialogue component
	- database component
	- error handling component.
	When test criteria have been defined for each component of a function, a test plan is created which is supported by the test scripts, the input test data requirements and expected output data results.
	These items should be placed under Configuration Management control, so that a function may be easily re-tested should it change.
	Additionally, all functions, particularly the on-line ones, must be assessed for their conformance to standards described in the parts of the customer's Installation Style Guide which were included in the OR, and/or in the Application Style Guide, agreed between the customer and supplier.
	At the time of performing the tests against the physical system, the logical products can be mapped to their physical counterparts by means of the traceability paths.
Testing I/O components	An I/O component represents the data content of a function in terms of data attributes that are entered (input) and/or attributes that are reported (output). Acceptance criteria must be specified to ensure that this

I/O data is present, has the expected data value and is of the correct type and format.

The SSADM products that may be used to derive the test criteria of an I/O component are:

- I/O Structures
- Screen Formats
- Report Formats
- Data Catalogue.

Testing dialogue components

A dialogue component relates primarily to on-line functions and to off-line functions with an on-line element, and represents the possible pathways through a dialogue. Criteria must be established to ensure that the navigation of a screen is acceptable, and that, where appropriate, the function may call other related functions. It is also necessary to ensure that only the appropriate users may gain access to the function, and that all the points in the system from which a function may be called are considered.

The SSADM products that may be used to derive the test criteria of a dialogue component are:

- Dialogue Structures
- Dialogue Control Tables
- Command Structures
- User Role/Function matrix.

Testing database components

A database component is concerned with the database accesses which a function has to make.

Each event or enquiry within a function should be considered separately, in order to test that it accesses the correct entities in the required way (for example, Create, Modify, Delete or Read) and that the integrity of the database is intact after the event or enquiry is complete.

It may also be considered part of the testing of this component to ensure that, where restrictions on data access by a user(s) exist, an entity may only be updated by those users with the correct access rights. The Entity Descriptions are the source for identifying individual user access restrictions.

Protection of the integrity of the data also needs to be considered, ie tests must be made to ensure that the system prevents an update process amending an entity, if the entity occurrence is at an inappropriate state in its life history. The 'main' source of data integrity requirements are the ELHs of the entities.

The SSADM products that may be used to derive the test criteria of database components are:

- ECDs, EAPS or Process Models
- Event Entity Matrix
- ELHs
- Entity Descriptions.

The test scripts should ensure that all selections and iterations on the ECDs, EAPS and/or Process Models are tested.

Testing error-handling components

An error-handling component relates primarily to on-line functions and to off-line functions with an on-line element. The acceptance tests must ensure that errors are reported correctly, concisely and unambiguously.

The error-handling components support the preceding components, in that they recognize errors, report the condition and take the appropriate further action. Whether test scripts for error-handling components should be specified separately or as a part of the scripts for other component types depends on the technical platform on which the system has been implemented.

Performance tests

This type of acceptance testing is concerned with evaluating the performance of the system functionality in meeting the business requirements. There are two forms:

- normal loading tests, which attempt to emulate the performance of the function when the system and the hardware on which it will run are running under the 'normal' load

- stress loading, which tests the behaviour of the function under extreme circumstances, eg when the system is operating at maximum capacity.

For both forms of test the performance of the functions is compared with all the relevant non-functional requirements. These may be function-specific, or they may be non-functional requirements which apply to all or many functions.

The creation of such test conditions by the customer provides an opportunity to measure some of the non-functional requirements which apply to the system as a whole, eg the time to recover from a machine failure. Section 10.3.2 describes this type of testing.

It should be remembered that performance tests are based on estimates of data volumetrics and loadings, and the customer's ability to successfully emulate that loading on the system. The ability of the system to meet the required service levels for each function should be continually monitored. Provision for this should be part of the system's Service Level Agreement (SLA).

10.3.2 Testing system-level non-functional requirements

These requirements encompass such aspects as system availability, service support and technical environment constraints.

The SSADM products that may be used to derive the test criteria for these types of requirement are:

- The Requirements Catalogue

- The Technical Environment Description (TED).

Technical environment requirements are recorded in the first instance as design constraints in the Requirements Catalogue. They may be amplified or clarified in the TED, which should be updated by the supplier to provide a description of the physical system.

Thus, the TED becomes a base-line against which the delivered environment is measured and any hardware commissioned.

The TED may also document the solution for some of the non-functional requirements which apply to the functions, particularly those relating to system security, eg password protection and secure communication lines.

There are likely to be other forms of system-level requirement which relate to the level of service which the customer requires of the system. These requirements are not normally traceable to SSADM products. Their solution is resolved by the creation of a Service Level Agreement (SLA) between the customer and supplier.

VOLUME STRUCTURE

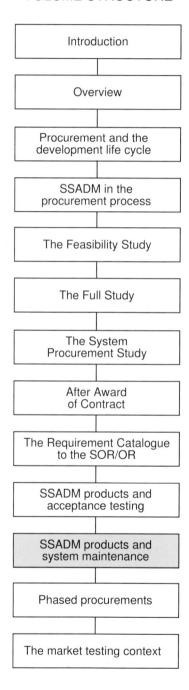

11 SSADM products and system maintenance

The drafting of an implementation contract must take account of the maintenance of the system after it has gone into operation. The implementation contract must include provisions which either stipulate the design products to be maintained by the supplier, including the SSADM design products, or, if the customer is procuring the design and source code in order to maintain the system himself, it must stipulate the design products to be made available to the customer. This chapter discusses some of these issues.

The return on investment of an SSADM Requirements Specification and of an SSADM design are greatly increased if they are used to maintain the system during its operational life. This additional purpose for the SSADM products is unaffected by whether the system is maintained by the customer or by the supplier. However, not all of the SSADM specification products and not all of the SSADM design products are essential for carrying out maintenance effectively. This chapter identifies the SSADM products which are most useful.

11.1 Contracted-out and in-house support

At the time of evaluating Full Proposals the customer must consider the suppliers' proposals concerning maintenance of the system. If an SSADM design is being procured, one of the evaluation criteria should be the extent to which the supplier proposes that maintenance of the system should be based on SSADM products.

The implementation contract agreed with the chosen supplier must include provisions for the maintenance of the system after it has gone into operation. The provisions vary, depending on whether the customer takes responsibility for the maintenance of the system, or whether the supplier maintains the system as part of the contract with the customer.

The most common scenarios for the maintenance of information systems is summarized in Figure 11.1.

The three scenarios in the figure represent in overview the most typical permutations of implementation and system maintenance contracts. For the purposes of this

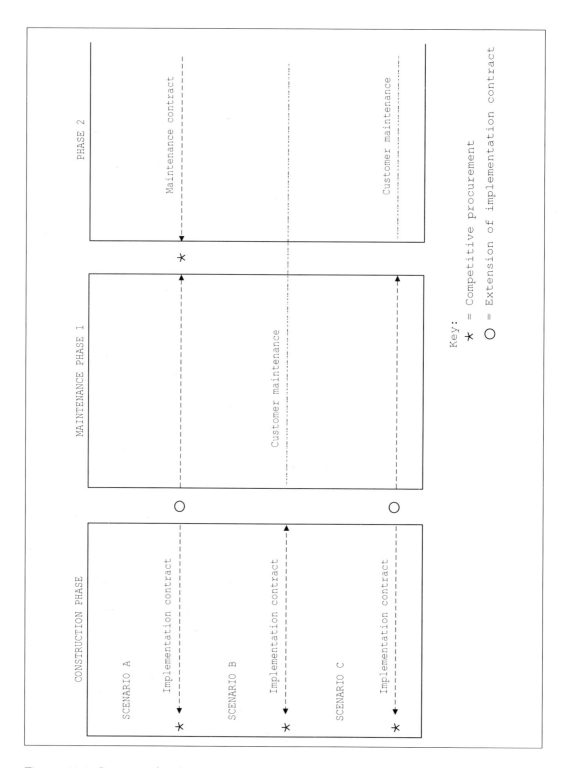

Figure 11.1: Contracts for the maintenance of the system - the most common options

section, each of the scenarios is presumed to start with a competitive procurement.

In scenario A, the customer undertakes a competitive procurement, which results in the award of an implementation contract. The Full Proposals received at this point include the suppliers' proposals concerning:

- which SSADM design products will be developed

- which SSADM products will be maintained and used in the maintenance of the operational system.

The customer awards the implementation contract to the supplier who, among other criteria, has proposed the most appropriate set of SSADM design products for the proposed development environment.

Prior to the operation of the system, the customer undertakes a review – represented as 'O' on the figure – of the supplier's performance against the implementation contract, and decides whether or not to continue with the supplier into the maintenance phase – 'Maintenance Phase 1'. The figure shows the case where the customer is satisfied with the supplier's performance, and the implementation contract continues as the basis for the supplier's activities during the first maintenance phase. (The options open to the customer if he is not satisfied with the supplier's performance are not shown.)

At the end of Scenario A the initial implementation contract expires, and the customer undertakes another competitive procurement which results in the award of a maintenance contract to a supplier. This supplier may either be the original supplier, or a third party.

In Scenario B, the customer wishes to maintain the system after it has been accepted for operational use. Therefore, the initial competitive procurement results in the award of an implementation contract which covers only the construction of the system, not its maintenance. In this scenario the Full Proposals include the suppliers' proposals concerning:

- which SSADM design products will be developed

- which SSADM products the supplier suggests should be used by the customer in the maintenance of the operational system

- the format in which the design products will be supplied to the customer, for example, magnetic medium output from a CASE tool.

As in Scenario A, the customer awards the implementation contract to the supplier who, among other criteria, has proposed the most appropriate set of SSADM design products for the proposed technical environment. With regard to this aspect of the evaluation, the criteria are the same whether it is the customer or the supplier who will be responsible for system maintenance.

Scenario C does not introduce any further issues. It is included only to cover the rather less commonplace scenario, in which the customer awards an implementation contract including initial maintenance of the system by the supplier, but in which, after the initial contract expires, the customer takes over maintenance of the system (rather than undertaking a competitive procurement for the maintenance).

The evaluation criteria which the customer applies to the suppliers' Full Proposals are exactly the same in this scenario as in the two other scenarios. The questions the customer asks are in essence:

- 'Taking the proposed technical environments into account, which supplier is proposing the most helpful set of SSADM design products with which maintenance of my system can be controlled?'

- 'Could my staff effectively maintain the system with the products proposed and with appropriate training/technical support?'

In each of the three scenarios the suppliers include in their Full Proposals:

- the specification of the SSADM products of Stage 6 Physical Design and, where appropriate, those of

Stage 5 Logical Design from which they propose to develop them

- an indication of which SSADM design products would be developed under the contract but not subsequently maintained

- an indication of which SSADM design products would be both developed and used in the maintenance environment

- the reasons for not developing SSADM design products omitted from the proposal.

During negotiation of the Draft Contracts the customer reaches agreement with the supplier about the actual SSADM design products which will be developed and maintained. These negotiations may result in additions to or subtractions from the suppliers' initial proposals. It is also normally the case that during such negotiations the customer seeks clarification from the supplier about the purpose, derivation and format of the proposed design products.

11.2 SSADM products in the maintenance environment

This section gives guidance on the SSADM products which are likely to be most useful in the maintenance environment. For the purposes of this section it is assumed that the system to be maintained has been wholly or substantially derived from an SSADM logical and physical design.

Not all of the suggested products will be relevant to all systems, and the format and content of the suggested design products will vary according to the chosen supplier's development environment. Customers should reach their own conclusions about which products will be helpful, drawing on relevant technical advice where necessary.

11.2.1 Maintenance and requirements traceability

It is of great benefit in the maintenance environment to have the requirements traceability paths available in the form defined by the analysts and designers:

- the paths are more likely to be complete and accurate, because they are put in place as each of the SSADM products is prepared rather than as a once-and-for-all exercise

- the paths may be used to identify all the products affected by a change request – this helps in the preparation of accurate resource estimates for maintenance activities

- time and effort on planning re-tests of the system is greatly reduced, because test paths and products required to test the paths are already documented

- those who join the maintenance team after the system has gone into operation are able to find out why the system has been designed the way it is, by tracing back to the business objectives and requirements addressed by the system.

Chapter 10 describes the requirements traceability paths supported by SSADM.

11.2.2 SSADM products used in maintenance

There are two types of SSADM product needed to maintain an information system derived wholly or substantially from an SSADM design:

- SSADM specification products

- SSADM design products.

The SSADM specification products are used by the customer as the means by which required enhancements to the system are expressed. It is a matter for the customer's Configuration Management system to control the point at which accepted enhancement requests are reflected in the documentation.

The SSADM specification products are used by the maintenance team, to gain an understanding of what the system is required to do, and what the quality criteria are which apply to the system. This is very important to new members of the maintenance team, who will find overview products, such as the Data Flow Model and the Technical Environment Description, particularly helpful.

The SSADM design products provide the detailed technical description of the system which the team needs, in order to locate physical components of the system which are to be modified, and to design and implement the modifications.

SSADM PRODUCTS IN THE MAINTENANCE ENVIRONMENT	
SPECIFICATION PRODUCTS	DESIGN PRODUCTS
Requirements Catalogue	Application Development Standards
Logical Data Model	Physical Data Design
Data Catalogue	Function Component Implementation Map
Function Definitions	Process Data Interface
Entity Life Histories	
Effect Correspondence Diagrams or Update Process Models	
Enquiry Access Paths or Enquiry Process Models	
Technical Environment Description	

Overview Products:
    Data Flow Model
    User Role/Function Matrix
    Event Entity Matrix	

USEFUL TO:
 	
    Users
<--->
 	
                        Maintenance Team
<--->	

Table 11.1

The balance of this chapter addresses each column of Table 11.1 in turn, describing in more detail each of the entries in the table and their relationship.

Column 1 – SSADM Specification Products

There are significant advantages to the customer, if users continue to 'own' the SSADM Requirements Specification, and if enhancements to the system are expressed in the terms of SSADM specification products with which the users became familiar during the Full Study, instead of in the terms of physical products with which the supplier has implemented the system.

Put another way, there is a risk to the customer that if changes to the system are expressed only in terms of physical products, then over time the user's understanding of the requirement may become limited by the way in which the supplier has implemented the solution.

The first group in the column headed 'Specification Products' contains those products needed by the customer to express changes to the business requirement, eg an additional data item, or a change in a business rule such as an algorithm. During the operational life of every information system other changes are made to the system which are not the result of changes to the business requirement but are the result of shortcomings in the physical design, eg amendments to the size of success units, or rationalisation of common processing.

The second group of specification products provides overviews of the system from the viewpoint of:

- Processes
- User Roles
- Functions
- Entities
- Events.

These products are not normally used to express system enhancements. For example, a change in the processing of an event is best expressed by means of an Effect Correspondence Diagram or an Update Process Model rather than by means of the Event Entity Matrix.

Requirements Catalogue	This should continue to act as the central repository of all the requirements. A well drafted Requirements Catalogue is likely to remain the SSADM product most easily understood by the users, including users brought into the maintenance environment after the system has gone into operation.

Each entry in the Requirements Catalogue should be annotated with its 'resolution'(s). There may be more than one resolution per entry, each cross-referring either to another specification product which amplifies or clarifies the entry, or to a physical design product or physical component of the operational system which implements the entry. |
| Logical Data Model (LDM) | The LDM models the business data in terms of the system's entities (ie data groups) and the relationships between them. The LDM is an important element of the conceptual model of the system.

All three elements of the LDM should be maintained:

- The Logical Data Structure
- The Entity Descriptions
- The Relationship Descriptions.

If the requirements traceability paths in the SSADM products have been kept as simple as possible, the information normally recorded in Relationship Descriptions will be recorded in the Entity Descriptions. |
| Data Catalogue | The Data Catalogue provides the detailed description of the data items in the information system. In the maintenance environment it is likely that the Data Catalogue is maintained in the form of (a component of) a Data Dictionary System, or equivalent.

By the time the system goes into operation, the validation tests and error messages associated with each data item should be included in the Data Catalogue (or its equivalent). |

Function Definitions	Function Definitions should document the user view of the logical system processing.

The function documentation should consist of:

- Function Definitions

- I/O Structures supported by I/O Structure Descriptions – though I/O Structures which have been superseded by Dialogue Structures should not be maintained

- Dialogue Structures (for on-line functions) supported by Dialogue Element Descriptions and Dialogue Control Tables

- Command Structures (for on-line functions)

- Screen Formats

- Report Formats. |
| User Roles | User Roles describe the categories of user with on-line access to the system, and the activities on the system which each category of user is able to undertake. |
| Effect Correspondence Diagrams (ECDs)

Enquiry Access Paths (EAPs)

Update Process Models (UPMs)

Enquiry Process Models (EPMs) | In SSADM these products are the building bricks, as it were, of the conceptual model of the system. They document the system processing at the level of the events and enquiries.

In the maintenance environment the following should be maintained:

   either
      Effect Correspondence Diagrams (for events) and
      Enquiry Access Paths (for enquiries)

   or, if available,
      Update Process Models (for events) and Enquiry
      Process Models (for enquiries).

Note that the value of maintaining ECDs and EAPs is greatly diminished if UPMs and EPMs are available. In order to reduce the maintenance overhead, ECDs and |

	EAPs may be dropped if UPMs and EPMs are being maintained.
Entity Life Histories (ELHs)	ELHs chart all of the events that may cause an entity to be changed. In SSADM the ELHs are an important part of the conceptual model of the system, and provide newcomers to the maintenance team with an easy-to-understand description of the system processing.

If it is felt that maintaining the operations on the ELHs is too much of an overhead, the diagrams should continue to be maintained without the operations. The detrimental effect of not maintaining operations on the ELHs is reduced where Update Process Models are available. |
| Technical Environment Description (TED) | The TED describes the functional and physical aspects of the system as it has been implemented, such as:

- the actual hardware and software, including their physical location
- the telecommunications hardware and software
- the fall-back and recovery arrangements
- disaster recovery arrangements
- the system's access and security
- system sizing, such as file sizes.

This information is essential in the maintenance environment, and should be kept up-to-date at all times. |
| (Top Level) Data Flow Model (DFM) | If the required system DFM is maintained during the procurement project, it should continue to be maintained in the maintenance environment.

The DFM is not a precise specification product. It is, therefore, not suitable as a means of describing required enhancements to the system. In the maintenance environment the purpose of the DFM is to act as an introduction to the system for those who need an understanding of its scope, eg managers of other |

projects, and analysts and programmers brought into the maintenance team.

The Function Definitions, and the SSADM products which support them, are more appropriate to the detailed description of the system's processing than a DFM. Therefore, it is sensible to maintain only a top level DFM. It should consist of:

- an un-decomposed Data Flow Diagram

- Elementary Process Descriptions, which summarize the processes formerly described at a lower level of decomposition

- I/O Descriptions, at least in narrative form, and, if possible, listing the (most important) business data items on data flows crossing the system boundary

- External Entity Descriptions describing the other systems with which the system interfaces, and the users which have access to the system.

User Role/Function Matrix

The User Role/Function Matrix provides an overview of the on-line functionality of the system in terms of:

- what functions are available on-line

- who is able to access the on-line functions

- who should not be able to access the on-line functions.

In short, it provides an at-a-glance summary of the whole of the on-line functionality of the system at the function level.

Event Entity Matrix

In SSADM the Event Entity Matrix identifies which entities are affected by a particular event, and what the nature of the update is in terms of 'create', 'modify' or 'delete'. It thus provides an at-a-glance overview of the on-line and off-line functionality of the system at the event level.

If 'entities' are added to the matrix to represent system files, which are not derived from the LDM's entities (eg look-up files), and if the enquiries are added to the matrix, and if 'read only' is inserted in the grid where appropriate against both events and enquiries, then the matrix provides an at-a-glance summary of all functionality of the system at the event/enquiry level, both on-line and off-line.

If it is known which physical files implement the entities, and if it is known which physical processes implement the events and enquiries, then the matrix is particularly helpful when physical files are changed, because the matrix shows all of the processes affected by the change.

Column 2 – SSADM Design Products

Application Development Standards

The Application Development Standards define the standards which apply to the physical design and to the maintenance activities.

The parts of the Application Development Standards which should be maintained are:

- Application Style Guide
- Application Naming Standards
- Physical Design Strategy.

Physical Data Design

In SSADM the Physical Data Design is an essential part of the internal design of the system. It is the implementation specific design for the physical database or its equivalent, if the solution does not include a Database Management System.

The Physical Data Design must be maintained, in order for changes to the physical data files to be managed and controlled.

Function Component Implementation Map (FCIM)

The FCIM bridges the gap between the Function Definitions and the physically implemented system. It does this in two ways:

- it specifies the components of functions not specified elsewhere, eg control and error handling, physical dialogues, input/output routines such as sorts and merges, syntax error handling

- it provides the cross-reference between the components of the functions and the physical objects which implement the components.

The more complex the implemented system, the more helpful the FCIM becomes. For example, where the system is an integration of physical design elements which have been produced by more than one supplier on a phased procurement, or where the system is a mixture of package and bespoke software, it is the FCIM which documents how the physical software components fit together to implement the required functionality.

Process Data Interface (PDI)

The PDI documents the physical database access fragments which implement the logical database accesses specified for the update and enquiry processes.

One of the main benefits of the PDI is that it enables the user to continue viewing the implemented database as the LDM, ie it enables the user to express change requests to the system processing in terms of logical, business-oriented products instead of in terms of physical products.

Another important benefit of the PDI is that it enables the maintenance team to continue using the UPMs and the EPMs as the principal specifications of the physical processes. In a 4GL the UPMs and EPMs may be the only products which specify the physical processes.

Note, however, that the PDI cannot be developed unless UPMs and EPMs have been developed. Therefore, the PDI should only be maintained where UPMs and EPMs are also being maintained.

Chapter 11
SSADM products and system maintenance

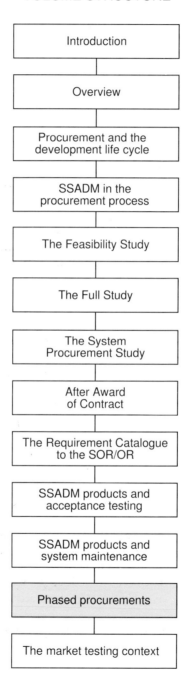

12 Phased procurements

The focus in the other chapters is upon the use of SSADM with a 'big bang' procurement strategy. The purpose of this chapter is to consider the effect upon the use of SSADM in a phased procurement strategy.

12.1 The purpose of phased procurements

Implementing a major information system in one 'big bang' is a high risk procurement and implementation strategy. Phased implementation normally offers a more manageable and lower risk option but does need careful planning and management control.

A phased procurement is one in which the customer divides the required system into phases and procures the phases separately. The phases may be procured serially, so that one phase is complete before the procurement of the next phase begins, or they may be procured with varying amounts of overlap, for example, Phase 2 begins when Phase 1 has been designed and built but not implemented.

When deciding upon a procurement strategy, there are two overriding considerations:

- risk to the achievement of business objectives from
    - technical risks
    - project management risks
    - cost risks
    - risks associated with managing change in the user environment

- evolving requirements.

If either of these two considerations features strongly in the project, then a phased procurement strategy should be considered. A phased procurement strategy has the objectives of reducing implementation risks and costs over the life of the project to a more manageable level.

However, a phased procurement strategy carries its own risks:

- the overall timescale will be longer, possibly reducing the value of the expected benefits

- if the phases overlap, there is a risk that decisions made in a later phase may be prejudiced by events in a preceding phase or invalidated by events which subsequently occur in another phase. Careful planning and management are required to control this risk.

12.2 The impact of phasing on SSADM

Exactly what is procured in the phases of a phased procurement varies from project to project and depends on the nature of the requirement, the resources available and other business factors. A phased procurement can only have an impact on the use of SSADM where the customer procures the design of the system in phases.

Thus if, for example, the customer procures a network-based infrastructure in Phase 1 and an application system in Phase 2, then there is no impact on the use of SSADM since all of the design activity falls into a single phase. But if the customer procures both the infrastructure and the application system in Phase 1 and procures an extension to that application system in Phase 2, then there is an impact on the use of SSADM since the design of the solution straddles both phases.

Figure 12.1 summarizes the impact of a phased procurement upon SSADM. It is intended to highlight the way in which, on a phased procurement, certain activities happen once only while others are revisited in each phase.

The figure shows the stages of the SSADM life cycle as a vertical column of boxes and distinguishes the stages covered by SSADM from those not covered. The loop from the stage at which a phase is implemented – box 9 in the figure – to the Business System Option selection for a phase – decision point C on the figure – represents the completion and acceptance of one phase and the commencement of the next phase.

The procurement process is summarized in the figure as one box – see box 5. In order to keep the figure simple most of the procurement transactions between the

Chapter 12
Phased procurements

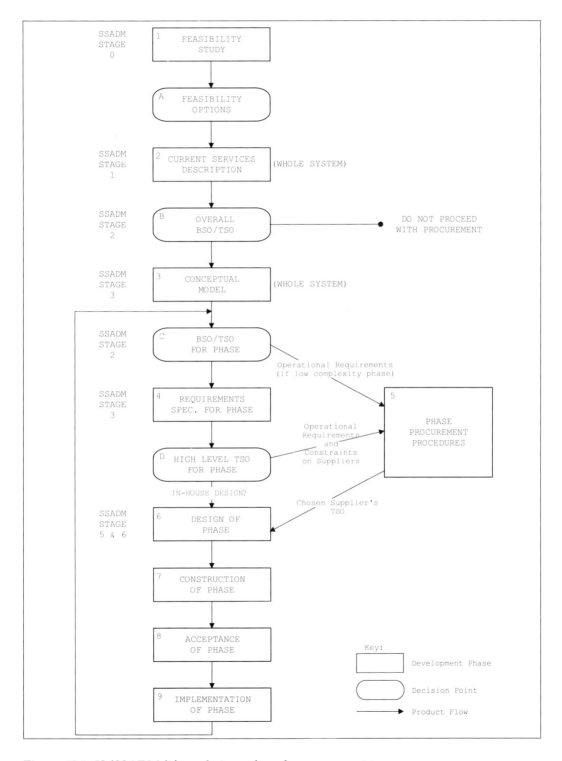

Figure 12.1: IS/SSADM life cycle in a phased procurement

customer and the supplier, such as, proposals and draft contracts, have been omitted. However, in order to highlight the flow around the loop the input of the Operational Requirement into the procurement process and of the chosen supplier's Technical System Option into the design stage are shown.

When to decide upon a phased procurement

At the time of deciding upon the procurement strategy, whether in the SSADM Stage 0 Feasibility Study or in SSADM Stage 2 Business System Options, the customer should consider the arguments between a 'big bang' procurement and a phased procurement strategy. In considering the latter, the customer should also consider the candidate phases and their relationships.

The decision to adopt a phased procurement strategy may be made when selecting either the Feasibility Option (decision point A in the figure) or the Business System Option (decision point B in the figure).

Although the decision to adopt a phased procurement strategy is taken for business and management reasons, risk reduction and/or the spreading of costs, the decision must take into account Technical System Option issues, in order to minimize possible integration problems. Technical interfaces needed to enable system growth from one phase to another must also be carefully defined.

Revisiting BSOs and TSOs

In a phased procurement the Business and Technical System Options are developed at two levels. Before the first phase the BSOs and, if appropriate, the TSOs are determined for the overall system. Then for each phase these options are reviewed, revised and augmented to take account of experience to date, changing circumstances and the more detailed matters which are better decided at the time of the phase procurement rather than at the outset.

The adoption of a phased procurement strategy does not mean that Business System Option selection can be divided into phases. It is as important on a phased procurement project, as on any other type of project, to consider those factors of the overall system which are always taken into account during BSO selection. These include the costs and benefits of the system as a whole, the impact on working practices, changes to the customer organization, the functional requirements, the non-functional requirements and the inter-dependencies between these factors.

Also, as with any other type of project, it may be necessary to take high-level Technical System Option decisions at the time of selecting the overall Business System Option. For example, it may be decided that all of the phased implementations must communicate with one another by means of the customer's existing network.

This overall BSO selection is represented in the figure by decision point B.

When the system boundary has been put in place, as a result of the overall BSO selection, the conceptual model of the system as a whole should be developed as represented by box 3 in the figure. The importance of developing the conceptual model in advance of Phase 1 is discussed below.

For each phase of the procurement there may be BSO/TSO decisions to be taken which are specific to the phase, though for Phase 1 the issues are likely to have been considered as part of the overall BSO/TSO exercise. For example, customer management may wish to revisit the costs and benefits associated with a phase; it may be necessary to allocate new mandatory/desirable priorities to the phase-specific requirements; the preceding phase may have had an unforeseen impact upon the requirement. Decision point C in the figure represents this activity.

As with a 'big bang' procurement, there may be TSO decisions at each phase of the procurement which are only identified during detailed analysis of the requirement, such as whether the required forms are to be 120 or 132 characters in width. This is indicated by decision point D on the figure.

It is important that any BSO/TSO decisions taken in connection with a phase should be compatible with the constraints of the overall option selection. This is particularly true of system boundary issues, where adjustments to suit one phase may have contractual implications for a preceding phase.

Note that the path in the figure from decision point D to box 6, the design stage, allows for the possibility of developing a phase in-house.

The importance of a complete conceptual model

The successful management and control of separate design activities depends, in part, upon having a conceptual model of the whole system. The importance of the conceptual model as a management tool increases in proportion to the amount of bespoke design being undertaken by suppliers.

At the time of dividing the system and its design into phases the customer cannot know how many suppliers will be involved in producing the design. In the case of overlapping phases the customer may be faced with more than one supplier designing separate parts of the system at the same time.

In SSADM the conceptual model of a system consists of:

- the Logical Data Model
- the Entity Life Histories
- the Effect Correspondence Diagrams and, if available, the Update Process Models derived from them
- the Enquiry Access Paths and, if available, the Enquiry Process Models derived from them.

The importance of having a complete conceptual model available before the first procurement phase becomes clear if the impact of having only a partial model is considered.

If, for example, the customer specifies a partial LDM (containing some entities but not all) and a sub-set of events and enquiries which cover just that part of the system to be implemented in Phase 1, then the database implemented in Phase 1 will be derived from this partial model.

When the events and enquiries in Phase 2 are developed, however, it is highly likely that they will need the Phase 1 LDM to be modified:

- additional entities may be inserted between existing Phase 1 entities

- additional attributes may be required for the Phase 1 entities

- there may be different ways of accessing the Phase 1 entities.

Therefore, the customer is likely to be faced with great difficulties. If a separate Phase 2 database is developed, which in part duplicates the Phase 1 database, this introduces the problem of maintaining data integrity. Alternatively the customer does away with the Phase 1 database in order to develop an integrated data design. This introduces the problems of redesigning and retesting the Phase 1 processes and converting the Phase 1 data to the Phase 2 database, using the revised data structures.

Having considered and selected the Business System Options for the system as a whole, the customer should develop the SSADM products of the conceptual model. That is to say, the SSADM products listed above should be developed as fully as possible, together with the data requirements and the functional and non-functional requirements which apply to the system as a whole. Only when the conceptual model is understood should the customer commence activities which are specific to the first phase. (In practice, the BSO decisions for the overall system will probably incorporate those for the first phase).

However, developing the conceptual model at this early stage presents some difficulties. In terms of SSADM stages the development of the conceptual model is carried out in Stage 3 Definition of Requirements. It is standard SSADM practice to develop events and enquiries after the initial set of functions have been defined and to validate the LDM against the processing requirements of the events and enquiries. To develop the events, enquiries and LDM before the functions have been defined is to diverge from this practice.

There are two ways of approaching this. 'First-cut' Function Definitions and I/O Structures may be developed which contain only the information necessary to identify events and enquiries. Otherwise (as in SSADM version 3) the events may be identified from updates to data stores on the Data Flow Diagrams, without reference to functions, and the enquiries may be identified from the Requirements Catalogue.

Separate requirements specification for each phase

Each phase of the procurement should commence with consideration of the phase-specific BSO/TSO issues. The decisions taken on these issues, as with any other form of SSADM project, provide the objectives for the phase Requirements Specification. The preparation of this product is shown as box 4 on the diagram.

In terms of SSADM stages the Requirements Specification is developed in Stage 3. This stage is thus revisited once per phase. In each phase the objectives of Stage 3 are as follows:

- to specify the external design requirements – the function and dialogue requirements

- to review and validate the conceptual model

- to specify the interface requirements resulting from previous phases.

On a phased procurement a Statement of Requirement (possibly) and an Operational Requirement are prepared for each phase. The SSADM products which should be incorporated into the SOR and the OR are those described for a 'big bang' procurement in Chapter 6.

For the sake of simplicity, the figure shows SSADM Stage 5 (Logical Design) and Stage 6 (Physical Design) together in box 6. In other words it assumes that the customer procures the logical design of each phase from the supplier. However, as discussed in Chapter 6, the customer may complete parts or all of Stage 5 himself and include the SSADM logical design products in the Operational Requirement.

Knock-on and knock-back effects between phases

Knock-on effects are those affecting the domain of a later phase as a result of events in an earlier phase. Knock-back effects are those affecting the domain of an earlier phase as a result of events in a later phase. The Project Plan must provide for both customer and supplier resources to deal with knock-on and knock-back effects.

The design, build and implementation of separate phases is made easier if each phase is self-contained, especially in respect of the acceptance criteria. However, with the possible exception of procurements based on off-the-shelf packages, it is unlikely that the design can be divided into phases in such a way that knock-on and knock-back effects are entirely avoided.

Knock-on effects mostly have an impact on the suppliers. They constrain the design of a later phase. For example, the design of Phase 2 may need to accommodate data in the format presented by Phase 1. Developing such an interface requires supplier resources, which are likely to result in increased purchase costs for the customer. To minimize the impact of these knock-on effects, the interface requirements between the various phases should be included in the OR.

Knock-back effects have an impact on both the customer and the suppliers. They may arise during the specification of a later phase by the customer. For example, during the specification of Phase 2 mistakes may be identified in the parts of the conceptual model from which the internal design of Phase 1 was developed. In this case the customer is responsible for amending the conceptual model products and the supplier may have to amend the design of Phase 1, and rebuild it.

Knock-back effects may also occur where the later phase develops an integrated extension to the existing design, rather than bolting a (more or less) stand-alone design 'onto the side'. For example, physical data design in

Phase 2 may reveal inadequacies in the Logical Data Model. In this case the customer should amend the conceptual model; ideally the Phase 1 supplier should also amend the Phase 1 physical design.

Drawing the design together

If SSADM designs are used to build the solution to a phased procurement, then the customer will reap the following benefits:

- a full set of requirements traceability paths provided in the SSADM products, as described in Chapter 10

- a set of SSADM products with which maintenance and enhancement of the design can be carried out, as described in Chapter 11.

The extent of such benefits will depend on the extent that SSADM designs are employed throughout the phased procurement.

If a full set of requirements traceability paths is available, then it should be as easy to put together acceptance tests in later phases, when an integrated design may be tested, as it is for Phase 1. For example, when testing whether a function meets all of the non-functional requirements which apply to it, it is irrelevant to the requirements traceability path whether the function is part of Phase 1 or a subsequent phase.

If the appropriate SSADM specification products are prepared and maintained by the customer and if the suppliers prepare appropriate SSADM design products, then one avoids the problems of maintaining a system built with multiple design standards.

For example, one of the maintenance products should be the Function Component Implementation Map (FCIM). The FCIM is developed in parts; each part is developed by the supplier responsible for implementing the relevant functions. However, when complete, the FCIM documents all of the implemented functions and can provide the maintenance team with a view of the whole design. This can help identify, for instance, re-useable processes and screens, which might otherwise be duplicated.

Chapter 12
Phased procurements

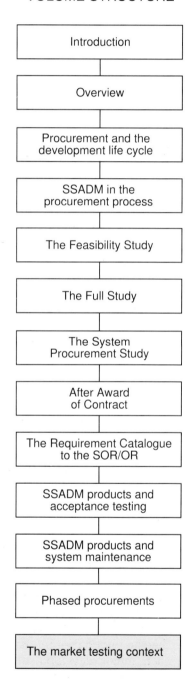

13 The market testing context

13.1 The impact on system development

If the organization's IT services have been market tested, and more significantly if they are outsourced, there will be a number of additional factors that project managers need to take into consideration. This section reviews the variety of circumstances that can apply in respect of systems development. Subsequent sections in this chapter consider the implications for new systems procurement and systems support.

Where there is still an in-house systems development capability, the guidance in the previous chapters will be fully applicable. This will also apply where the IT services are provided by the in-house team following a successful bid in a market test. In this latter case a more formal relationship is likely, between the system development unit and their customers, than was previously the case.

As a result of market testing the provision of some or all IT services, including systems development, may be contracted out. Even should all systems development be contracted out, the customer must retain the ability to act as an 'intelligent customer' of its system providers.

A number of options exist for the development of the requirements and the sourcing of the design and development of new systems, which are described in Section 13.2. However, in all these options, the specification of requirements for a new system is the responsibility of the user, and the design of the new system is the responsibility of the system supplier.

A distinction needs to be drawn between maintenance and the minor development of systems on the one hand, and major new systems or enhancements on the other hand. Under the IT services outsourcing contract, maintenance and minor development work is likely to be contracted to the outsourced IT service provider. The contract should also include a definition of what constitutes 'minor' development work. This work will then be undertaken by the IT services contractor. For such 'minor' developments the requirement specification should still be agreed by the user, together with the cost

203

of the work and implementation timescales, before the project proceeds to the design and implementation stages. SSADM will be used where this has been agreed with the contractor, usually as part of the IT services contract.

Of course, such arrangements need not always be the result of market testing. Market tests take many forms and may result in separate contracts, with separate contractors, for each of the elements of IT services.

In the case of major new systems and major enhancements, users should normally have the freedom to go out to competitive tender. This is a safeguard to ensure value-for-money from the project. The IT services contractor will normally have the right to tender in the competitive procurement. Bearing in mind the responsibility of the user for the specification of the requirement, it is desirable that this is performed under a separate contract, and separate contractor, from that for the design, supply and support of the new system. This may not always be possible, but where one contractor performs both the specification of the requirement, and the design, supply and support of the system, the risks of this practice must be taken into account. If the procurement of the requirements specification is over the EC/GATT threshold, the use of SSADM for this work cannot normally be mandated, but can be a desirable requirement.

13.2 New systems procurement

Under an outsourced IT services arrangement, a number of options exist for the supply of new systems, that need to be taken into consideration during the BSO and TSO stages. As discussed in Section 13.1, major new systems will normally be subject to a competitive procurement. Dependent on the circumstances, the client (that is the purchasing) organization might also have a choice as to who manages and operates the new system. These are factors that need to be considered in the feasibility and full studies.

There are three possibilities, as to who manages and operates the new system:

- it can be the existing IT services provider, under an extension to the IT services contract

- it can be the customer. This might apply if the new system is implemented on PCs or mini computers

- it can be another IT services contractor, under a new IT services contract.

The choice will depend primarily on the IT services contract. If not mandated there, the choice will depend on the technical quality of the alternative proposals, and the cost of the proposals over the life of the project.

In the first option above, even if the IT services contractor is not supplying the new system, their involvement will be required in three main areas. Firstly, they need to provide an input on the technical environment. Secondly, the cost of the new system that needs to be evaluated will be the cost in terms of the additional service charges tendered by the IT services contractor, not just the costs of software and hardware from third-party suppliers. Thirdly, they will at a minimum be closely involved in the system acceptance trials, and may take full responsibility for the conduct of those trials.

In the second and third options above, the IT services contractor does not necessarily need to be involved in the procurement, though the IT services contractor may be the customer's main source of technical advice.

13.3 Responsibilities for system support

Whenever a system is procured, arrangements need to be made for system support, and subsequent enhancement. Typically system support will be provided by the system supplier, though the IT services provider is a possible alternative. Whoever maintains the system, appropriate safeguards need to be built into the support contract.

Under an IT services outsourcing contract, whether system support is provided by the IT services contractor, or the system supplier, it is likely that support for the service will be provided via the IT service contractor's Help Desk. Whoever provides the support, it is important that the technical and cost implications are fully assessed during the procurement. For bespoke systems, the documentation required for the support is part of that assessment.

If the system documentation is to be to SSADM standards, that needs to be specified in the contract. Cost and procurement considerations will determine whether SSADM continues to be the preferred documentation for system support, when the system is procured externally. A major enhancement to a system previously using SSADM documentation, will raise different considerations from those applicable to a new system, since in the former case compatibility with existing system documentation could well be a pertinent cost and procurement issue.

Bibliography

This volume refers to the following CCTA products:

Appraisal and Evaluation Library

The Appraisal and Evaluation Library volumes are available from HMSO Bookshops and from HMSO Publications Centre, PO Box 276, London, SW8 5DT, telephone 071 873 9090 or fax 071 873 8200.

- Overview and Procedures
  ISBN: 0 11 330534 6

Information Systems Engineering Library

The Information Systems Engineering Library volumes are available from HMSO Bookshops and from HMSO Publications Centre, PO Box 276, London, SW8 5DT, telephone 071 873 9090 or fax 071 873 8200.

- SSADM and Application Packages
  ISBN: 0 11 330626 1

- Estimating using MK II Function Point Analysis
  ISBN: 0 11 330578 8

Information Systems Guides

The Information Systems Guides are available from John Wiley and Sons Ltd, Customer Services, Shripney Road, Bognor Regis, West Sussex, PO22 9SA or telephone 0243 829121

- CCTA IS Guide volume B4: Appraisal Investment in Information Systems
  ISBN: 0 47192529 2

Information Systems Procurement

The Information Systems Procurement volumes are available from HMSO Bookshops and from HMSO Publications Centre, PO Box 276, London, SW8 5DT, telephone 071 873 9090 or fax 071 873 8200.

- A Guide to Procurement within the Total Acquisition Process
  ISBN: 0 94668358 1

SSADM The set of 4 SSADM Version 4 Manuals are available from NCC Blackwell, Oxford House, Oxford Road, Manchester M1 7ED

- SSADM Version 4 Reference Manuals
  ISBN: 1 85554 004 5

Information on all CCTA products can be obtained through the CCTA Library, Rosebery Court, St Andrew's Business Park, Norwich, NR7 0HS. Tel. 0603-704704.

Glossary

acceptance trial	A trial carried out to determine whether a system delivered by a contractor meets a Department's requirements as specified in the contract.
Business Case	A document justifying the proposed expenditure on an IS Strategy or project in terms of business benefits.
Business System Option(BSO)	BSOs are the means by which users agree the new application's desired functionality with developers. BSOs are used to define the functionality needs and the boundary for the system, with reference to the business needs.
call-off contract	A contract which stipulates minimum (and usually maximum) quantities of goods and which commits the purchaser to place orders for at least the minimum quantity. The orders may be placed at any time during the duration of the contract subject to the terms and conditions governing ordering and delivery.
CCTA	The Government Centre for Information Systems
contractor	A supplier who has been awarded a contract.
CRAMM	CCTA Risk Assessment and Management Method
Data Flow Model	A set of Data Flow Diagrams which show how services are organised and processing is undertaken. It can act as an effective means of communication between analysts and users.
Draft Contract	A draft form of contract which is issued to shortlisted suppliers following evaluation of full proposals. The draft contract is then negotiated in parallel with technical discussions and demonstrations, leading to its agreement and finalisation. The agreed draft contract then forms the basis of a best and final offer.
EC Journal	Supplement to the Official Journal of the European Community; details available through HMSO, 51 Nine Elms Lane, London, SW8 5DR. Tel. 071-211 5656

Euromethod	Euromethod will be a framework within which existing IS methods can work. It will provide a high level description of the processes by which customers and suppliers should interact during the development of Information Systems.
evaluation	The task of measuring, against predetermined criteria, a supplier's ability to meet the procuring Department's requirements stated in a Statement of Requirement, or an Operational Requirement.
facilities management	An arrangement whereby a contractor manages, operates and supports all, or part of, a client's IT services. Facilities management services can be provided in a variety of ways to meet the differing requirements of client organisations.
feasibility study	A short assessment of an information system proposed in the IS strategy.
Full Proposal	A document forming a step in the procurement process, in which a supplier responds to an Operational Requirement.
full study	An early and major component of the system development life cycle, which analyses present systems (if any), specifies user requirements in detail, assesses ways of meeting those requirements (through the Mini-Proposal process when a procurement is involved) and produces a Business Case for the proposed system.
Function Component Implementation Map	A classification and specification of all implementation fragments for all function components defined in the Function Definitions to meet the processing requirements.
Function Definitions	The packaging of all details about functions to be included in the Requirements Specification.
GATT	General Agreement on Tariffs and Trade.
information system	Any procedure or process, manual or IT-based, which provides a way of acquiring, storing, processing or disseminating information.

Invitation To Tender (ITT)	A document forming a step in the Direct to ITT process, containing a description of a particular supplier's proposed solution to the procuring Department's requirements, and designed to draw from that supplier a tender setting out precise commitments on costing, timescales etc.
IS Strategy	A formal definition of the intended future deployment of IS, and its supporting policies, programmes of work, and infrastructure, within a Department in support of business objectives.
Logical Data Model (LDM)	Provides an accurate model of the information requirements of all or part of an organisation. It serves as a basis for file and database design, but is independent of any specific implementation technique or product.
Mini-Proposal	A document forming a step in the procurement process, with which a supplier responds with his solution to a Statement of Requirement.
Operational Requirement (OR)	A document forming a step in the procurement process, containing a complete statement of the procuring Department's requirements, addressed to one or more potential suppliers of equipment or services, and designed to draw from each supplier a Full Proposal describing in detail how the supplier could meet the requirements.
post implementation review	A formal review to determine the extent to which a completed project has met its objectives and, in particular, has achieved the expected benefits.
PRINCE	Projects in Controlled Environments, the standard methodology for project management used for IS projects in UK Government Departments.
quality review	A formal procedure in which a product of a project is checked against an agreed set of quality criteria.
requirement	A functional, operational or performance need, specified in the business or operational terms to the user Department.

Requirements Catalogue	Is the central repository for information covering all identified requirements, both functional and non-functional.
Requirements Specification	A packaging of all the details which are required in order to decide upon the technical direction of a project.
solution	The IT system or sub-system, and associated services, proposed by a supplier to meet one or more requirements.
SSADM	Structured System Analysis and Design Methodology, a standard methodology for analysis and design in software development projects, developed by CCTA and used in UK Government Departments and in the private sector.
Statement of Requirement (SOR)	A document produced at an early stage in the procurement process, which specifies the core or mandatory requirements of the Department procuring the information system. This document is addressed to suppliers, and they are invited to respond with Mini-Proposals.
supplier	A commercial company or other organization which offers to supply goods or services to Government in response to requests for proposals or tenders. When contracts have been awarded, the successful suppliers are referred to as 'contractors'.
system procurement study (SPS)	A part or fully funded exercise, in which two or more suppliers each carry out a detailed study, to develop their solutions in response to the requirements of a high complexity IS project. The purpose of an SPS is to reduce the risks involved in system implementation.
TAP	Total Acquisition Process. Developed by CCTA, TAP provides guidance on good practice and a formal set of procedures for the effective management of the processes involved in the acquisition of Information Systems and related services.
Technical System Option (TSO)	TSOs are the means by which users agree a new application's implementation strategy incorporating the desired functionality, as defined in the Requirements

	Specification. several TSOs are developed and one is selected (or combined from several). this gives the technical direction for future development.
Update Process Model (UPM)	A structure diagram for the update (event) processing and the associated operations list.

Index

3-schema specification architecture 34
3GL 94, 113, 114
4GL 38, 49, 94, 114, 188
Acceptance Trial 130, 149
BSO (*see* Business System Options)
Business Case 23, 48, 65, 88, 151
Business System Options (BSO) 15, 25, 29, 58-60, 62, 66, 67, 69, 77, 78, 80, 84, 86, 89, 91, 95, 121, 131, 146, 192, 194, 195, 197, 198, 204
CASE tool 132, 178
CCTA 9, 50, 53, 58, 75, 122
conceptual model 93, 183-185, 195-200
contractor(s) 55, 203-205
cost 13, 14, 23, 26, 29, 47, 50, 62, 75, 81, 105, 110, 124, 191, 203, 205, 206
CRAMM 13
Data Flow Diagram(s) (DFD) 79-81, 89, 121, 122, 186, 198
Data Flow Model 27, 54, 77, 112, 120, 143, 180, 181, 185
Define Required System Processing 27, 68
Definition of Requirements 15, 23, 60, 67, 77, 106, 197
Develop Required Data Model 27, 68
DFD (*see* Data Flow Diagram)
EAP (*see* Enquiry Access Path)
ECD (*see* Effect Correspondence Diagram)
Effect Correspondence Diagram(s) (ECD) 74, 75, 90, 93, 97, 161, 163, 171, 181, 182, 184, 196
ELH (*see* Entity Life History)
Enquiry Access Path(s) (EAP) 74, 75, 90, 93, 161, 163, 171, 181, 184, 185, 196
Enquiry Process Model (s) 74, 90, 93, 94, 110, 112, 114, 161, 163, 181, 184, 185, 188, 196
Entity Life History(ies) 74, 90, 92-94, 96, 161, 163, 171, 181, 184, 185, 188, 196
EPM (*see* Enquiry Process Model)
Euromethod 10, 15
External Design 56, 102, 109, 113, 198
facilities management 40
Feasibility Option(s) 27, 47-56, 59, 60, 62, 63, 66, 146, 194
Feasibility Report 140
Feasibility Study 10, 25, 27, 29, 42, 47-49, 51-53, 58, 59, 61, 62, 65-67, 131, 146, 194

215

Full Study 10, 27, 30, 42, 44, 47, 50, 53, 57-62, 65-70, 76, 87-89, 96-98, 102, 103, 105, 106, 108, 112, 121, 123, 149, 151, 162, 165, 167, 182
Function Component Implementation Map 43, 112, 116, 118, 163, 164, 181, 187, 200
Function Definition(s) 89, 91, 92, 96-99, 100, 102, 118, 158, 159, 162-164, 168, 181, 184, 186, 187, 198
GATT 9, 34, 47, 50, 51, 53, 57, 58, 86, 127, 204
Graphical User Interface (GUI) 102
GUI (*see* Graphical User Interface)
High Complexity 23, 31, 45, 60, 65, 105, 109
Information System 9, 11, 18, 29, 43, 48, 54, 56, 59, 61, 63, 65, 67, 78, 90, 92, 95, 101, 135, 140, 141, 152, 166, 180, 182, 183, 191
Internal Design 56, 69, 98, 187, 199
Investigation of Current Environment 15, 29, 58-61, 66
investment appraisal 58, 59
IS Strategy 23, 25, 48-50, 65, 140
ITT 67
Jackson 41, 99
LDM (*see* Logical Data Model)
LDS (*see* Logical Data Structure)
Logical Data Model(s) (LDM) 27, 36, 43, 54, 74, 77, 78, 81, 82, 85, 87, 89-92, 117, 132, 139, 143, 144, 163, 164, 181, 183, 187, 188, 196-198, 200
Logical Data Structure (LDS) 75, 82, 83, 183
Logical Design 15, 16, 29, 31, 69, 97, 101, 106, 108, 129, 179, 198, 199
Low Complexity 58, 63, 67, 87, 88
Medium Complexity 60, 61, 68, 76, 87-89
menu 91, 103, 112, 114
Mini-Proposal 23, 26, 27, 30, 76, 77
Operational Requirement (OR) 9-11, 13, 14, 16-18, 23, 25-31, 34, 37-43, 45, 47-61, 65-72, 74-79, 82-103, 105, 106, 108-120, 122-124, 127-132, 135-146, 149, 151, 157-159, 161-165, 167-172, 175, 177-184, 186-188, 191, 192, 194, 195, 198-200, 203, 205
OR (*see* Operational Requirement)
package-based 28, 34, 36, 39, 52
package-constrained 28, 29, 34, 37, 38, 54, 63
PDI (*see* Process Data Interface)
phased procurement(s) 11, 47, 67, 109, 188, 191-194, 198, 200

Physical Design 16, 17, 28-31, 43, 70, 72, 106, 108-112, 114, 115, 118, 123, 129, 130, 153, 155, 178, 179, 182, 183, 187, 188, 198, 200
physical environment 112, 115, 119, 120
physical implementation 115, 132, 166
PIR (*see* Post-Implementation Review)
Post-Implementation Review (PIR) 131
PRINCE 13
Process Data Interface (PDI) 43, 110, 116, 162, 164, 181, 188
Project initiation 131
project plan(s) 51, 199
Requirements Catalogue 11, 16, 36, 38, 42-44, 50, 51, 54, 66, 67, 71, 72, 77-80, 83-87, 90, 92, 95, 97, 100, 102, 103, 112, 120, 131, 135-138, 140-143, 145, 155, 157-160, 162, 172, 181, 183, 198
Requirements Specification 13, 14, 16, 17, 26, 27, 30, 56, 57, 94, 113, 120, 124, 128, 129, 132, 135, 151-153, 167, 168, 175, 182, 198, 204
Required System Logical Data Model 27, 74, 77
security requirements 160
SOR (*see* Statement of Requirement)
Specification of Requirements 203
SPS (*see* System Procurement Study)
Stage 0 15, 27, 42, 47, 50, 51, 62, 194
Stage 1 15, 29, 58-62, 66, 80, 87
Stage 2 15, 25, 29, 58-60, 62, 66, 67, 131, 194
Stage 3 15, 23, 25-27, 30, 37, 60, 67-70, 77, 101, 102, 106, 123, 197, 198
Stage 4 14, 15, 21, 23, 25, 30, 68
Stage 5 15, 16, 27-31, 69, 70, 101, 106, 108-110, 112, 123, 124, 179, 198, 199
Stage 6 16, 17, 28-30, 97, 106, 108-112, 114, 116-118, 123, 162, 178, 198
Statement of Requirement (SOR) 11, 14, 16-18, 23, 26-28, 42, 43, 51, 60, 61, 63, 65-68, 70, 71, 76-89, 92, 94-96, 127, 135-138, 140-143, 145, 146, 198
Step(s) 27, 68, 129, 131
Step 210 131
Step 310 68
Step 320 27, 68
System Procurement Study (SPS) 11, 26, 27, 31, 45, 94, 101, 105-115, 117-120, 122-124
TAP (see Total Acquisition Process)

Technical Environment Description (TED) 78, 79, 85, 86, 90, 94, 100, 103, 160, 161, 165, 166, 172, 173, 180, 181, 185
Technical System Options 14, 23, 25, 30, 36, 68, 69, 91, 103, 128, 194, 195, 198, 204
TED (*see* Technical Environment Description)
Total Acquisition Process (TAP) 9, 11, 23, 31, 45, 50, 53, 59, 60, 65, 67, 76, 87-89, 105
TSO (*see* Technical System Options)
Update Process Model(s) (UPM) 28, 74, 90, 93, 94, 97, 110, 112, 114, 161, 163, 181, 182, 184, 185, 188, 196
user role(s) 78, 82, 86, 87, 91, 96, 100, 113, 114, 131, 158, 160, 162, 170, 181, 182, 184, 186